Advances in Design Methods from Modeling Languages for Embedded Systems and SoC's

Lecture Notes in Electrical Engineering

Volume 63

Advances in Design Methods from Modeling Languages
for Embedded Systems and SoC's
Dominique Borrione
ISBN 978-90-481-9303-5, 2010

Advances in Communication Systems and Electrical Engineering
Xu Huang, Yuh-Shyan Chen, and Sio-Iong Ao
ISBN 978-0-387-74937-2, 2008

Time-Domain Beamforming and Blind Source Separation
Julien Bourgeois, and Wolfgang Minker
ISBN 978-0-387-68835-0, 2007

Digital Noise Monitoring of Defect Origin
Telman Aliev
ISBN 978-0-387-71753-1, 2007

Multi-Carrier Spread Spectrum 2007
Simon Plass, Armin Dammann, Stefan Kaiser, and K. Fazel
ISBN 978-1-4020-6128-8, 2007

For other titles published in this series, go to
http://www.springer.com/series/7818

Dominique Borrione
Editor

Advances in Design Methods from Modeling Languages for Embedded Systems and SoC's

Selected Contributions on Specification, Design, and Verification from FDL 2009

Springer

Editor
Dominique Borrione
TIMA Laboratory
(CNRS, Grenoble INP, UJF)
Avenue Felix Viallet 46
38031 Grenoble
France
dominique.borrione@imag.fr

ISSN 1876-1100 e-ISSN 1876-1119
ISBN 978-90-481-9303-5 e-ISBN 978-90-481-9304-2
DOI 10.1007/978-90-481-9304-2
Springer Dordrecht Heidelberg London New York

Library of Congress Control Number: 2010934420

Cover design: SPi Publisher Services

Printed on acid-free paper

Springer is part of Springer Science+Business Media (www.springer.com)

Preface

More than ever, FDL is the place for researchers, developers, industry designers, academia, and EDA tool companies to present and to learn about the latest scientific achievements, practical applications and users experiences in the domain of specification and design languages. FDL covers the modeling and design methods, and their latest supporting tools, for complex embedded systems, systems on chip, and heterogeneous systems.

FDL 2009 is the 12th in a series of events that were held in Lausanne, Lyon, Tübingen, Marseille, Frankfurt am main, Lille, Barcelona and Stuttgart. In 2009, FDL was organized in the attractive south of France area of Sophia Antipolis, together with the DASIP (Design and Architectures for Signal and Image Processing) Conference and the SAME (Sophia Antipolis MicroElectronics) Forum.

All submitted papers were carefully reviewed to build a program with 27 full and 10 short contributions. From these, the Program Committee selected a shorter list, based on the evaluations of the reviewers, and the originality and relevance of the work that was presented at the Forum. The revised, and sometimes extended versions of these contributions constitute the chapters of this volume.

The four parts of this book reflect the organization of FDL 2009 around four Technical Areas. Each one of them is preceded by an introduction by the Technical Area Chair.

Part I, UML and MDE for Embedded System, addresses model driven methods, based on the Unified Modeling Language, for performance analysis, validation and verification. The two chapters involve the use of the MARTE (Modeling and Analysis of Real-Time and Embedded systems) UML profile, standardized in late 2009.

Part II, C/C++-Based System Design, corresponds to the area that attracted the largest number of submissions, and contains six chapters. Most of them use SystemC, or extensions to SystemC, for modeling complex designs and performing exploration of architectural alternatives, performance evaluation, validation and synthesis. Yet, two chapters advocate the use of functional languages for those tasks.

Part III, Embedded Analog and Mixed-Signal System Design, includes two chapters that address the modeling and validation of heterogeneous and RF circuits, by means of dedicated simulators based on SystemC-AMS and VHDL-AMS.

Part IV, Assertion Based Design, Verification & Debug, contains four chapters that present new concepts and methods that take as input formal properties for verification, synthesis and stimuli generation. The use of assertions is no longer limited to the register transfer level, and is now made available for transaction-level models.

As editor of this book, I am indebted to the many persons who made it possible to appear. I would like to thank all the reviewers who evaluated the articles submitted to FDL 2009, and provided detailed and well-argued comments from which the best papers could be selected. The four Technical Area Chairs deserve special acknowledgement; they are, from Part I to Part IV: Pierre Boulet (Université Lille 1, France), Frank Oppenheimer (OFFIS, Oldenburg, Germany), Sorin A. Huss and Hans Eveking (both from Technical University of Darmstadt, Germany). Many thanks are due also to the authors, who worked hard to revise and improve their contributed article.

As a result this book presents extensions to standard specification and description languages, as well as new language-based design techniques and methodologies to solve the challenges raised by mixed signal and multiprocessor systems on a chip. It is intended as a reference for researchers and lecturers, as well as a state of the art milestone for designers and CAD developers. It covers a broad range of design objectives and modeling levels, and should be useful to the design and CAD community.

Dominique Borrione
FDL 2009 General Chair
Laboratoire TIMA (CNRS, Grenoble INP, UJF), France

Contents

Part I

UML AND MDE FOR EMBEDDED SYSTEMS

Model driven methods, mostly based on the Unified Modeling Language, increasingly support semi-formal methods for system level design of complex embedded systems, including highly programmable platforms and heterogeneous Systems-on-Chip. Current design methods do not close the gap from specification to (automatic) synthesis yet, but the two chapters of this part contribute toward this goal. Both build upon the modeling possibilities offered by the recent MARTE (Modeling and Analysis of Real-Time and Embedded systems) standard UML profile, whose version 1.0 has been made available in late 2009 to improve the productivity of design teams.

The first chapter, by Mallet, André and de Simone proposes to extend the capabilities of the IP-XACT standard, dedicated to the description of the interfaces of reusable blocks (the so-called Intellectual Properties), with a modeling of their timing properties using the MARTE standard UML profile. The combination of these standards improves the possibilities of reuse and analysis at (UML) model level, thus leading to an increase in design productivity and validation capabilities.

The second chapter, by Murillo, Mura and Prevostini presents a semi-automatic model based design flow using model driven engineering to interface several hardware–software co-design tools and to generate SystemC simulation code. This is made possible by the non-functional property specification capabilities introduced in the MARTE standard.

Pierre Boulet
FDL'09 UMES Technical Area Chair
Université Lille 1, France

Chapter 1

IP-XACT COMPONENTS WITH ABSTRACT TIME CHARACTERIZATION

Frédéric Mallet[1], Charles André[1], and Robert de Simone[2]

[1]*Aoste Team - I3S/INRIA, Université Nice Sophia Antipolis, 06902 Sophia Antipolis, France*
{Frederic.Mallet, Charles.Andre} @sophia.inria.fr

[2]*Aoste Team - I3S/INRIA, INRIA Sophia-Antipolis Méditerranée, 06902 Sophia Antipolis, France*
Robert.de_Simone@sophia.inria.fr

Abstract Large System-on-Chips are built by assembly of components modeled at different representation levels (TLM, RTL). The IP-Xact standard focuses on structure, type and memory information and ignores behavior and time issues. The UML profile for MARTE and its companion language CCSL provide advanced time modeling capabilities. By combining UML MARTE and IP-Xact, we introduce a more abstract timed representation level allowing the description of IP-Xact designs with UML-based environments. This paper discusses the use of MARTE to annotate IP-Xact specifications with time requirements. These time requirements are first used in simulation to generate waveforms. Then, actual implementations are considered and adequate observers are generated to validate these implementations with respect to the MARTE specification. The proposal is illustrated on the Leon2 architecture.

Keywords: UML, MARTE, IP-Xact, Time model, CCSL

1. Introduction

Reuse and integration of heterogeneous Intellectual Properties (IPs) from multiple vendors is a major issue of System-on-Chip (SoC) design. Existing tools attempt to validate assembled designs by global co-simulation at the implementation level. This fails more and more due to the increasing complexity and size of actual SoCs. Thus, there is a clear demand for a multi-level description of SoC, with verification, analysis and optimization possibly conducted at the various modeling levels. In particular, analysis of general

D. Borrione (ed.), *Advances in Design Methods from Modeling Languages for Embedded Systems and SoC's*, Lecture Notes in Electrical Engineering 63,
DOI 10.1007/978-90-481-9304-2_1,

platform partitioning based on a coarse abstraction of IP components is highly looked after. This requires interoperability of IP components described at the corresponding stages, and the use of traceability to switch between different abstraction layers. Although this is partially promoted by emerging standards, it is still insufficiently supported by current methodologies. Such standards include SystemC [8], IP-Xact [21], OpenAccess API [7], and also recent Unified Modeling Language (UML [14]) based standards like the UML Profile for Modeling and Analysis of Real-Time and Embedded systems (MARTE [15]) that specifically targets real-time and embedded systems.

System Modeling requires representation of both structural/architectural platform-based aspects at different levels of abstraction as well as behavioral/functional aspects possibly considering time-related viewpoints such as untimed/asynchronous/causal, logical synchronous/cycle-accurate or physical/timed models. Semantics provides behavioral meaning to the whole systems from the combination of the behavior of its components.

For *system structure* representation, UML uses class, component, and composite structure diagrams, while SysML [23] uses block diagrams. Tools like Esterel Studio, and virtual platforms like CoWare, Synopsys CoreAssembler and ARM RealView, introduce their own architecture diagrams. IP-Xact provides some ADL (Architecture Description Language) features for externally visible common interfaces and recognized state encodings, together with administrative information (ownership, tool chains, versioning...). IP-Xact has its own XML syntax, for specification of IP meta-data and tool interfaces.

For *component behavior* representation, SystemC provides programming libraries to represent IP component behavior at different abstraction levels, from Transaction Level Modeling (TLM) to Register Transfer Level (RTL) but it requires additional support for architecture modeling. Indeed, it does not provide yet support for the whole range of abstraction levels usually considered [4]. In that way, the commonly used SystemC levels, such as TLM/PV, TLM/CC, or RTL, are complementary to UML state-machine, sequence and activity diagrams that can be thought as closer to TLM/CP (Communicating Processes) level models. MARTE could be seen as introducing the relevant *timed* version at this level (like PVT does for PV - Programmer View), through logical time and abstract user-defined clock threads. However, MARTE is general-purpose and lacks some specific features of IP-Xact.

IP-Xact and SystemC provide little or no support to model at an abstraction level higher than TLM PV (Programmer View). Moreover, IP-Xact mostly deals with the structural aspects of IPs. The way IPs interact, their time characteristics or in general their behavior, are not covered by IP-Xact 1.5 which only rely on the associated SystemC/HDL code. However, early validation of IP-Xact design requires an abstract description of component behavior and of their time characteristics. The SPRINT Project (`http://www.sprint-project.net`), which aims at providing an extension of IP-Xact, highlights these needs.

Being a general-purpose modeling language, UML offers various diagrams (like activity, state machine or sequence) to represent the system *untimed* behavior. The MARTE profile, which includes the non-normative Clock Constraint Specification Language (CCSL) [10], extends the UML with explicit time-related concepts. Combined with UML behavioral elements it provides support for building models of the whole IP behavior at a *timed* Communicating Process (CPT) level.

By extending UML with IP-Xact specific features, we get a larger notion of component that combines the structural features of IP-Xact with timed UML behavior. Figure 1.1 shows this unified notion of component: on the left side, the structural aspects covered by IP-Xact; on the right side, the behavioral aspects brought by the UML (shown with a hatched background). The UML functional models can be annotated with time information using the MARTE time model. This defines the timed behavior of the whole IP. The relationship between the structural aspects and the behavior, which is implicit in IP-Xact is made explicit here. Writing in fields of control registers may trigger the execution of some behavior, which in turn uses values of other registers as input parameters.

In this unified environment we get structural, functional as well as time descriptions of IPs. We propose to go one step further by extending IP-Xact descriptions with time requirements of IPs. It is neither practical nor desirable to include the whole behavior since it would require adding all UML behavioral model elements to IP-Xact. It is also not practical in most cases since implementations of the same IP at different abstraction levels are usually made by different teams and may result in components that do not even have the same interfaces. For example, a simple read or write communication at the TLM level

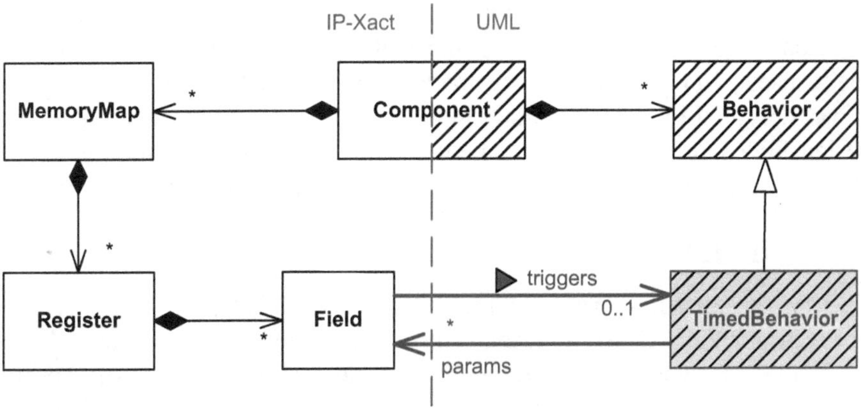

Figure 1.1. Unifying IP-Xact and UML components

implies several control signals at the RTL level. The same RTL signals can also be shared by completely different transactions. Rather than addressing the whole IP behavior, we focus on their time requirements. These requirements, missing in IP-Xact, are usually described as waveforms in datasheets. In Section 4, we specify them as a CCSL specification. Then, we rely on CCSL operational semantics to execute the specification and automatically produce waveforms. In a second phase (Section 5), we use this specification to assemble dedicated observers that are to be run in testbenches against candidate implementations both at the RTL and at the TLM levels.

Section 2 briefly discusses related work. Section 3 gives an overview of MARTE CCSL. The following sections describe our proposal to add time requirements to IP-Xact components and to use it to verify candidate implementations both at the RTL and TLM levels. The proposal is illustrated on a bridge extracted from the implementation of the Leon2 architecture.

2. Related Work

There have been several propositions to use UML in SoC Design [5, 19] including usage of profiling mechanisms (UML for SoC [11, 13], Omega-RT [6] and UML for SystemC [18]). There are also some combined UML/SysML-based environments to support analysis and produce SystemC outputs [22]. However, our work specifically focuses on the interoperability among IP-Xact models and makes an extensive use of the MARTE profile and its time model.

Some preliminary works [2, 20, 24] have started to consider solutions to model IP-Xact designs in general purpose modeling languages like UML with or without the support of MARTE profile. These approaches mostly focus on structural aspects, whereas we also consider behavior and time information of IPs.

The recent UML profile for ESL [16, 17] supports bidirectional transformations between UML and IP-Xact as well as the generation of SystemC code skeletons based on the register map information provided by IP-Xact. This profile focuses on TLM models and abstracts away all the RTL-related information. It was designed to provide a good integration with ST Microelectronics TLM design flow. Our work is complementary. The profile for ESL focuses on the structural aspects while we also look at the relationship with the behavior.

3. MARTE CCSL

The UML is a general-purpose language that can be adapted to domain-specific needs through its extension mechanisms. Its time model is simplistic and must be extended if to address real-time and embedded systems. The UML 2 profile for MARTE proposes a richer time model that supports physical/logical, discrete/dense time. It supersedes and extends the former UML 1.x Profile for

Schedulability, Performance and Time (SPT) [12]. Its stereotypes can be used at several levels: specification, analysis or design. At the specification level, time requirements specify the expected behavior of a component and clock constraints that must be enforced. CCSL is an OCL-like language, defined in an annex of MARTE, that proposes formally-defined time relations to constrain model elements. Any model element that should be constrained by CCSL must first be associated with a *time line*. Each time line is an ordered set of instants. Each instant denotes a moment in time at which the model element reacts. Depending on the model element the reaction may differ. Messages are sent or received, actions start or finish, behaviors are executed, operations are called or complete. These time lines, called *clocks* in MARTE, can represent any time event occurring in the model. They can refer to physical time, with so called *chronometric* clocks. *Logical clocks* refer to discrete-time logical clocks and represent logical time. Logical time is the time model in use in synchronous languages. Logical clocks as originally defined by Lamport [9] are a special case.

CCSL operators and their semantics have been previously defined [1]. We only recall here some operators required to understand our proposal. Most of the CCSL operators deal with discrete clocks (logical or chronometric) and dense clocks are discretized when to be combined with discrete clocks. CCSL rely on three basic relations on instants (occurrences of events): *coincidence*, *strict precedence* and *exclusion*. Clock relations combine these three basic operators to build relations on several (often infinitely many) instants of clocks.

Coincidence-based relations build subclocks or superclocks from a reference clock. We often use (infinite) periodic binary words to select the instants of the super clocks with which the instants of the subclock are coincident. For instance, $sub \boxed{=} \left(super \blacktriangledown 0^2.1^\omega\right)$ states that sub is a subclock of $super$, so that the first instant of sub is coincident with the third instant of $super$, the second with the fourth, and so on. The first two instants of $super$ are skipped.

Precedence-based relations build infinitely many instant precedence relations. For instance, $a \boxed{\prec} b$ states that the ith instant of a strictly precedes the ith instant of b, for all $i \in \mathbb{N}^\star$.

By combining both kinds, powerful relations can be derived. Other required constraints are introduced when used.

4. Example – AHB to APB Bridge

To illustrate our approach, we have considered the example of the Leon2-based platform architecture (Fig. 1.2) delivered together with the IP-Xact 1.4 specification. In this paper, the discussion mainly focuses on the APB bridge between the AHB (Advanced High-performance Bus) and the APB (Advanced Peripheral Bus) buses.

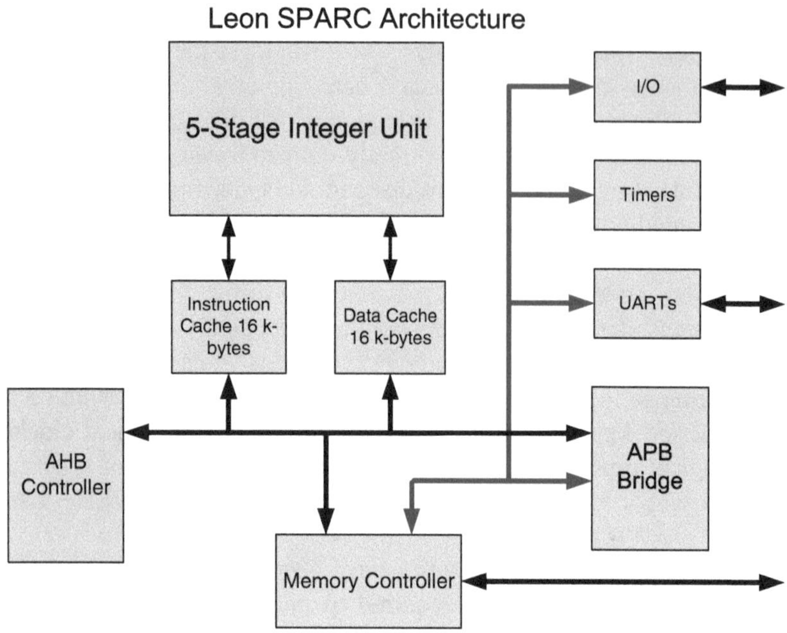

Figure 1.2. Leon 2 based platform architecture

The Leon2 architecture consists of IP-XACT components like processor, memory, bus arbiter, bus bridge, and timers. The various components are connected through two AMBA buses [3]: the AHB, when high throughput is required and the APB, for slower devices. We have considered, in particular, the communications between the processor and the timer. The processor is a master for the AHB and the communication must go through the bridge to reach the timer accessible only from the APB. After an informal description of the bridge we make this description formal with CCSL.

Informal Specification

Figure 1.3 shows a component description of the bridge. The bridge acts as a slave for the AHB (signals Hxxx) and as a master for the APB (signals Pxxx) and must therefore combine both protocols. The AHB arbiter selects (signal HSEL) the bridge when the communication must go through it. The bridge must in turn decode the address available from HADDR and select the right APB slave. There can be as many as 16 APB slaves. On write transactions, the APB bridge provides the select (PSEL), enable (PENABLE), write control (PWRITE), address (PADDR) and data (PWDATA) to the targeted peripheral. On read transactions, it multiplexes the targeted peripheral's data (PRDATA-*x*) to the AHB HRDATA with the proper timing.

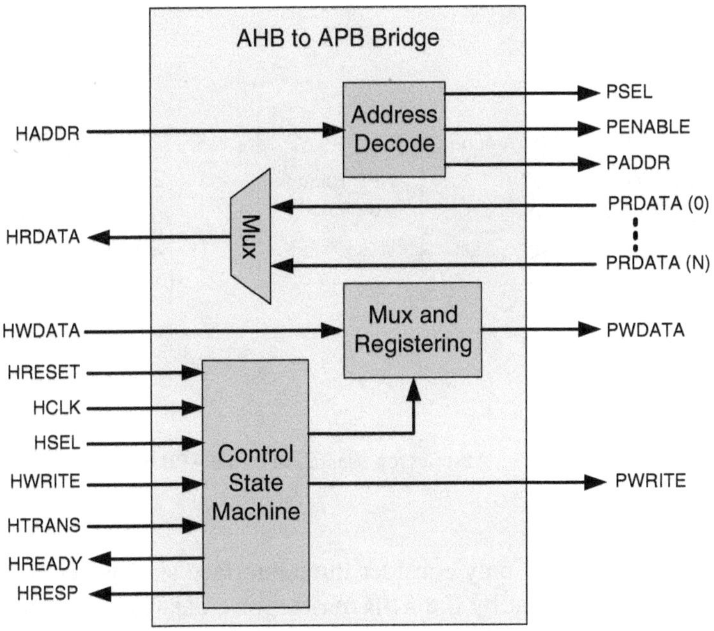

Figure 1.3. AHB to APB bridge

The bridge also returns the signal HREADY back to the AHB master to indicate that it has completed the APB transaction and the slave is ready for the next transaction. As per the AMBA Bus Specification, the APB bridge implementation must contain *two buffers* to store the address for the next transaction (read or write) while performing the actual one. The signal HREADY turns low when the buffer is full, telling the AHB master to stop sending more requests. When one operation is performed and one address can be discarded, the signal HREADY goes high again.

To illustrate, we have focused on this aspect of the communication. The AMBA specification states that two buffers are required and that two successive write operations can start without wait state. So, the APB bridge must behave like a bounded buffer of capacity 2 and the signal HREADY must be asserted only when the buffer is full, i.e. two addresses are stored. The following subsection uses CCSL to make this specification formal. Then, follows a discussion on the use of such a formal specification to compare different implementations of the bridge, possibly at different abstraction levels (TLM, RTL).

CCSL Specification

To make a formal specification of the behavior of the bridge, we start by building an abstraction that ignores the irrelevant signals. The selected abstraction

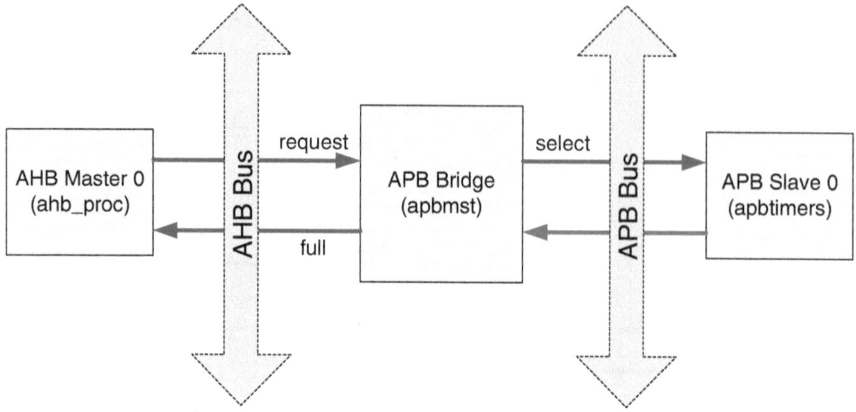

Figure 1.4. Abstract representation of the APB bridge

is shown in Fig. 1.4. We only consider three interface signals $request$, $select$ and $full$. $request$ is sent by the AHB master for a data transfer with an APB slave and is equivalent to the signal HSEL. After decoding, $select$ chooses the targeted peripheral, and forwards the request. This $select$ is an abstraction of the combined behavior of signals PENABLE and PSEL. To select a slave, the signal PSEL must be set to HIGH, then the rising edge on PENABLE triggers the actual selection. This combined behavior is abstracted as a single tick of the clock $select$. The signal $full$ is asserted when the buffer is full, which is roughly the negation of HREADY. HREADY is high when the bridge is available (not full) and goes low when it is not available. In our abstraction, $full$ ticks when the bridge is not available.

The APB bridge specification makes $request$, $select$, and $full$ interdependent. Firstly, the APB bridge must behave like a bounded buffer of capacity 2. Secondly, $full$ should be emitted whenever the buffer is getting full. The first requirement is a classical producer-consumer behavioral constraint. For any evolution, each occurrence of $select$ must be preceded by an occurrence of $request$, and the number of $request$ should not exceed the number of $select$ by more than 2. More formally:

$$(\forall i \in \mathbb{N}^{\star})\ request[i] \prec select[i] \prec request[i+2] \tag{1.1}$$

where $\prec$ is the precedence operator. The second requirement demands that $full$ occurs *as soon as* the condition $\#request = \#select + 2$ holds ($\#s$ denotes the number of occurrences of s, up to the observation instant). The *as-soon-as* operator entails a coincidence relation ($\equiv$) between an occurrence of *request* and an occurrence of *full*. These two requirements can be rigorously specified as clock constraints:

$$request \boxed{\prec} select \tag{1.2}$$

$$select \boxed{\prec} \left(request \blacktriangledown 0^2.1^\omega\right) \tag{1.3}$$

$$c1 \triangleq request \blacktriangledown 0.1^\omega \tag{1.4}$$

$$c2 \triangleq c1 \wedge select \tag{1.5}$$

$$full \boxed{=} c2 - select \tag{1.6}$$

The first two relations are the direct translation of Eq. 1.1. They use the precedence operator ($\boxed{\prec}$) on clocks. The semantics of relation 1.2 is precisely $(\forall i \in \mathbb{N}^\star)\ request[i] \prec select[i]$. Relation 1.3 is similar but applies on *select* and an implicit clock which is like *request* but deprived of its first two instants. $0^2.1^\omega$ is a periodic binary word that expands to the infinite binary word $0011 \cdots 1 \cdots$. The downward black triangle is the operator *filteredBy*. $c1$ and $c2$ are two local clocks introduced for clarity; parenthesized expressions could have been used instead. $c1$ is *request* without its first instant: $(\forall i \in \mathbb{N}^\star)\ c1[i] \equiv request[i+1]$. $c2$ is the inf ($\wedge$) of c1 and select, with respect to the precedence relation. The semantics of this relation is

$$(\forall i \in \mathbb{N}^\star)\ c2[i] \equiv \begin{cases} c1[i] & \text{if } c1[i] \preccurlyeq select[i], \\ select[i] & \text{otherwise.} \end{cases}$$

that is, $c2$ ticks in coincidence with the first ticking argument clock. When $c1$ is the "winner" of the race, the buffer is getting full. When the winner is *select*, the number of items in the buffer cannot increase, thus *full* must not tick. Eq. 1.6 expresses this contraint with a clock minus operator ($-$): *full* ticks only when $c1$ is the unique winner. Note that in the case of simultaneous ticks of $c1$ and *select*, *full* does not tick.

Figure 1.5 shows a simulation trace generated by TimeSquare, which is our dedicated environment to use MARTE time model and perform analysis of CCSL specifications (`http://www.inria.fr/sophia/aoste/dev/time_square`). Clocks can be displayed or hidden interactively. In the figure, only the non local clocks have been kept. The vertical lines between *full* pulses and *request* pulses are graphical annotations drawn by TimeSquare on the user's request and they highlight the coincidence between instants. The number of items (i.e. pending transfers) in the buffer is not part of the screen copy, it has been added for illustrative purpose. The behavior is as expected: *full* ticks when and only when the buffer enters the full state.

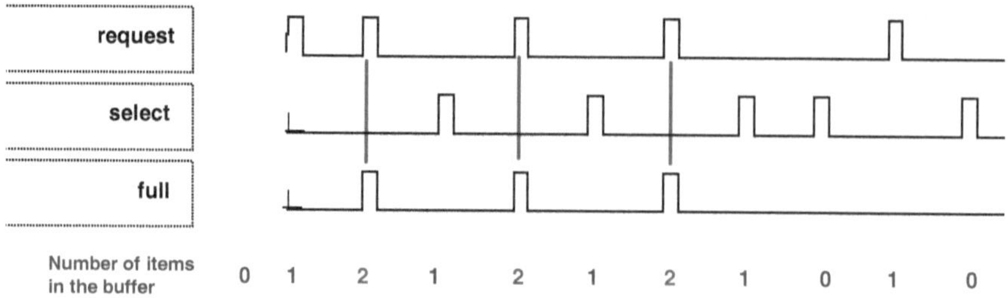

Figure 1.5. Simulation trace of constraints 1 to 5

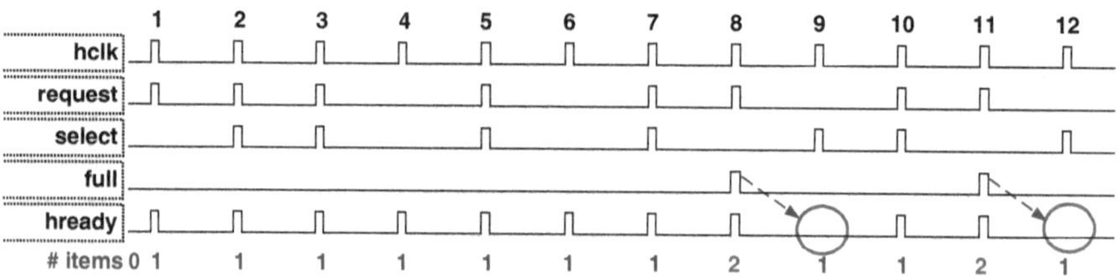

Figure 1.6. Synchronous Simulation trace

The APB bridge behavior is generally implemented in a synchronous way. The previous asynchronous specification can be refined into a synchronous one by introducing new (logical) clocks and imposing new clock constraints. Let $hclk$ be the clock of bus. The AMBA specification says that the $hready$ signal has to go low at the next $hclk$ tick following the saturation of the buffer. The textual CCSL specification below is appended to the previous one (Eqs. 1.2–1.6).

$$request \text{ isSubClockOf } hclk \tag{1.7}$$

$$select \text{ isSubClockOf } hclk \tag{1.8}$$

$$fullDlyd = full \text{ delayedFor } 1 \text{ on } hclk; \tag{1.9}$$

$$hready = hclk \text{ minus } fullDlyd; \tag{1.10}$$

Relations 1.7 and 1.8 force clocks $request$ and $select$ to be subclocks of $hclk$ (synchronous behavior). $fullDlyd$ is a local clock which is the already defined clock $full$ delayed for one period of $hclk$ (Eq. 1.9). The last relation (Eq. 1.10) makes the two clocks $hready$ and $fullDlyd$ complementary: when one ticks the other does not. This is illustrated in Fig. 1.6. The circles drawn at the 9th and 12th instants underline the absence of pulses of $hready$ at these instants, i.e. exactly one tick after the saturation of the buffer.

5. Comparing RTL and TLM Implementations

We use the formal specification defined in the previous section to verify that candidate implementations conform to this specification. We have used the implementations delivered to illustrate the IP-Xact 1.4 specification. Using the same CCSL implementation as a reference provides a way to compare different implementations of the same IP, even if they apply at different abstraction levels (RTL & TLM). Our proposal is to establish common time properties between the various implementations. This does not require the two implementations to have the same structure and can be applied to existing libraries of legacy IPs without imposing a common coding methodology to designers.

We start with a specification language (at CP level) that supports formal verification and code generation. At this level, we adorn the UML description of the component behavior with a CCSL specification of time properties to be verified (Fig. 1.7). For each CCSL relation, equivalent VHDL and SystemC observers have been defined. Observers are not intrusive and only

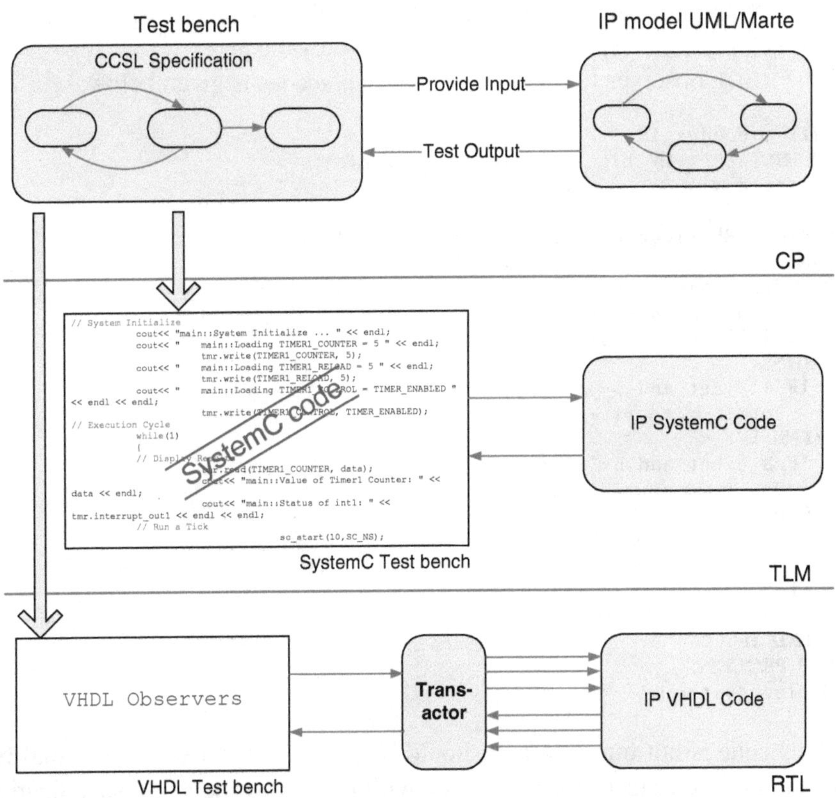

Figure 1.7. IP Test bench at various abstraction levels

observe the actual behavior of the implementation without interfering. When an unexpected behavior occurs, violation signals are asserted. A library of observers has been built to implement all CCSL constraints. The fact that these observers are actually conformant to the semantics of CCSL operators is not easy to prove and is out of the scope of this paper. However, most of the time, the implementation of a relation consists in the negation of the Boolean properties specified in the operational semantics of the relation.

In another work [1], the possibility to describe the observers in Esterel/SyncCharts has been studied with some applications to a different domain. The VHDL observers were automatically generated using Esterel Studio code generation facility. Here, we consider a library of VHDL/SystemC observers manually built. This discards the possibly to use Esterel formal verification environment but makes it easy to connect the observers to some actual implementations produced by other teams. Indeed, this may be very tricky with generated observers if the implementations do not follow some coding rules. Note that it is easier to build observers that detect unexpected behaviors than enforcing the correct behavior as TimeSquare does. Working with VHDL observers require to run a significant amount of testbenches to have a good coverage of the state space so a property can be validated.

The VHDL observer for the CCSL relation precedes is given below:

```
ENTITY precedes IS      -- a precedes b
   PORT (a,b : IN bit; violation : OUT bit);
END precedes;

ARCHITECTURE precedes_arch OF precedes IS
BEGIN
  PROCESS(a,b)
    -- count = #a - #b
    variable count : integer := 0;
  BEGIN
    IF a'event and a='1' THEN
      count := count + 1;
    END IF;
    IF b'event and b='1' then
      count := count - 1;
    END IF;
    IF count<0 THEN
      violation <= '1';  -- violation detected
    ELSE
      violation <= '0';
    END IF;
  END PROCESS;
END precedes_arch;
```

This code is not intrusive and should be run in parallel with the actual implementation to detect inconsistencies with respect to the CCSL specification. Another library of SystemC observers should be used to validate TLM implementations.

Since interfaces of the RTL and TLM models are not necessarily identical, the validation at the RTL (and possibly also at the TLM level) requires interface adapters, also called *transactors*. These transactors can often be generic. CCSL clocks do not directly represent the implementation signals. The CCSL specification is mainly event-based, while implementations may work indifferently on levels, on rising or falling edges. A great variety of adapters can be made available but there may always be a new one that must be built to adapt to the particular situation. For instance, the abstraction by the clock *select* of the combined behavior induced by signals PENABLE and PSEL requires a specific adaptor given below:

```
ENTITY adapter1 IS
    PORT (penable, psel : IN bit;
          cselect : OUT bit);
END adapter1;

ARCHITECTURE adapter1_arch OF adapter1 IS
BEGIN
  PROCESS(penable, psel)
  BEGIN
    IF psel'event and psel='1'
                  and penable='1' THEN
        cselect <= '1';
    ELSE
        cselect <= '0';
  END PROCESS;
END adapter1_arch;
```

Note that, these adapters are not bidirectional. In that precise example, the adapter adapts the VHDL signals of the available implementation and built an equivalent CCSL clock. Other adapters may do the contrary, *i.e.*, transform a CCSL clock to build equivalent VHDL signals.

We applied this technique systematically to build an observer for each clock relation. Our generated testbench acts as a passive component attached to the APB bridge. Violation signals are emitted during simulation and then collected for further analyses.

6. Conclusions

Early validation of systems built by assembling components requires to associate abstract behavioral and timing models with IPs. The recent SPIRIT IP-Xact 1.4 only considers structural aspects of the IPs. We consider here an integrated UML/IP-Xact framework in which the two standards are complementary. IP-Xact describes the structural part and transformations to UML structural diagrams can be built. UML brings the description of the behavior and combined with MARTE allows the specification of IP time properties. More precisely, we propose to combine the MARTE Time model and its constraint language CCSL with IP-Xact models. The CCSL description acts as a *formal* specification of time requirements for IP-Xact components as waveforms and timing diagrams do in the paper datasheets. The timing diagrams can be rebuilt from the CCSL description by using TimeSquare. Different VHDL or SystemC observers derived from the CCSL specification are then used to validate (through simulation) some properties of candidate implementations.

The proposal is illustrated on the time behavior of the Leon2 architecture distributed with the SPIRIT IP-Xact 1.4 specification. We specifically focus on the modeling and the validation of time properties related to communications that go through the AHB to APB bridge.

This contribution relies on simulation to validate properties. However, the formal semantics of CCSL opens perspectives to formal verification of IP-Xact models. This kind of extensions has successfully been applied to a signal processing application and must be explored for models in the domain of the Electronic Design Automation.

References

[1] C. André, F. Mallet, Specification and verification of time requirements with CCSL and esterel, in *LCTES*, ed. by C. Kirsch, M. Kandemir (ACM DL, SIGPLAN/SIGBED, Dublin, Ireland, 2009), pp. 167–176

[2] C. André, F. Mallet, A. Khan, R. de Simone, Modeling spirit IP-XACT in UML marte, in *Conf. on Design, Automation and Test in Europe (DATE), MARTE Workshop*, 2008, pp. 35–40

[3] ARM, *AMBA Specification.* ARM Limited. ARM IHI 0011A, 1999

[4] L. Cai, D. Gajski, Transaction level modeling: an overview, in *CODES+ISSS*, ed. by R. Gupta, Y. Nakamura, A. Orailoglu, P.H. Chou (ACM DL, Newport Beach, CA, USA, 2003), pp. 19–24

[5] R. Chen, M. Sgroi, L. Lavagno,G. Martin, A. Sangiovanni-Vincentelli, J. Rabaey, UML and platform-based design, in *UML for Real: Sesign of Embedded Real-Time Systems* (Kluwer, Norwell, MA, 2003), pp. 107–126

[6] S. Graf, I. Ober, I. Ober, A real-time profile for UML. *STTT* **8**(2), 113–127 (2006)

[7] M. Guiney, E. Leavitt An introduction to OpenAccess: an open source data model and API for IC design, in *ASP-DAC*, ed. by F. Hirose (IEEE, ACM DL, Yokohama, Japan, 2006), pp. 434–436

[8] IEEE Standards Association, *Open SystemC Language Reference Manual.* Open SystemC Initiative (IEEE Std. 2005), pp. 1666–2005

[9] L. Lamport, Time, clocks, and the ordering of events in a distributed system. Commun. ACM **21**(7), 558–565 (1978)

[10] F. Mallet, C. André, R. de Simone, CCSL: specifying clock constraints with UML/Marte *ISSE* **4**(3), 309–314 (2008)

[11] G. Martin, W. Mueller, *UML for SoC Design* (Springer, 2005), ISBN: 978-0-387-25744-0

[12] OMG *UML Profile for Schedulability, Performance, and Time Specification, v1.1*. Object Management Group, formal/05-01-02, 2005

[13] OMG *UML Profile for System on a Chip, v1.0.1*. Object Management Group, formal/06-08-01, 2006

[14] OMG *Unified Modeling Language, Superstructure*. Object Management Group, Version 2.1.2 formal/2007-11-02, 2007

[15] OMG *UML Profile for MARTE, v1.0*. Object Management Group, formal/2009-11-02, 2009

[16] S. Revol, S. Taha, F. Terrier, A. Clouard, S. Gérard, A. Radermacher, J.-L. Dekeyser Unifying HW analysis and SoC design flows by bridging two key standards: UML and IP-XACT, in *Distributed Embedded Systems: Design, Middleware and Resources*, volume 271 of *IFIP*, ed. by B. Kleinjohann, L. Kleinjohann, W. Wolf (Springer, 2008), pp. 69–78

[17] S. Revol, *Profil UML pour TLM: contribution à la formalisation et à l'automatisation du flot de conception et vérification des systèmes sur puce*. PhD thesis, INPG, Grenoble, France, 2008

[18] E. Riccobene, P. Scandurra, A. Rosti, S. Bocchio, A Soc design methodology involving a UML 2.0 profile for SystemC, in *Conference on Design, Automation and Test in Europe (DATE)* (IEEE Computer Society, Munich, Germany, 2005)

[19] T. Schattkowsky, UML 2.0 – overview and perspectives in SoC design, in *DATE* (IEEE Computer Society, Munich, Germany, 2005), pp. 832–833

[20] T. Schattkowsky, T. Xie, UML and IP-XACT for Integrated SPRINT IP Management, in *Design, Automation Conference (DAC), UML for SoC workshop*, 2008

[21] SPIRIT, *IP-XACT v1.4: A Specification for XML Meta-Data and Tool Interfaces*. Spirit Consortium, 2008, `http://www.spiritconsortium.org`.

[22] A. Viehl, T. Schönwald, O. Bringmann, W. Rosenstiel, Wolfgang, Formal performance analysis and simulation of UML/SysML models for ESL design, in *European Design and Automation Association*, ed. by G.G.E Gielen, *DATE* (Leuven, Belgium, 2006), pp. 242–247

[23] T. Weilkiens, *Systems Engineering with SysML/UML: Modeling, Analysis, Design* (The MK/OMG Press, Burlington, MA, 2008)

[24] J. Zimmermann, O. Bringmann, J. Gerlach, F. Schaefer, U. Nageldinger, Holistic system modeling and refinement of interconnected microelectronics systems, in *Conference on Design, Automation and Test in Europe (DATE), MARTE Workshop*, 2008

Chapter 2

MDE SUPPORT FOR HW/SW CODESIGN: A UML-BASED DESIGN FLOW

Luis Gabriel Murillo, Marcello Mura, and Mauro Prevostini
ALaRI Institute, Faculty of Informatics, University of Lugano, Lugano, Switzerland
{luis.gabriel.murillo, marcello.mura, mauro.prevostini} @usi.ch

Abstract The spread of technology is mostly possible thanks to the growing presence of Embedded Systems in our everyday lives. Embedded Systems are information processing devices enclosed in a variety of products such as cars, telecommunication or fabrication equipment. These devices must accomplish a large set of requirements and constraints to meet the computational demands of modern applications. As a matter of fact, Embedded Systems are very complex and heterogeneous systems, and this implies very difficult, expensive and time consuming design processes. Although MDE and Hw/Sw codesign are widely used to address the design complexity problem, the lack of design procedures and methodologies joining both concepts restrains their usage as complementary techniques, thus preventing the implementation of faster and more robust design cycles. In this chapter we present a practical semi-automated design flow where both methodologies are merged and exploited to enable a fast design process targeting highly complex Real-Time Embedded Systems, executing several tasks on SoC and MPSoC devices, while allowing the usage of Design Space Exploration, Schedulability Analysis and Estimation techniques.

Keywords: High level modeling, Model-driven engineering, Hardware/Software co-design, UML, MARTE, Design space exploration

1. Introduction

While fabrication technologies for integrated circuits evolve towards deep submicron levels, the semiconductor industry faces new design and development challenges mainly as a consequence of technology scaling and the increase in transistors number. This has a strong impact on system complexity that conflicts with the necessity for quick development of applications. Lately there has been a trend towards packaging more components in a single chip (i.e. SoC,

D. Borrione (ed.), *Advances in Design Methods from Modeling Languages for Embedded Systems and SoC's*, Lecture Notes in Electrical Engineering 63,
DOI 10.1007/978-90-481-9304-2_2, © Springer Science+Business Media B.V. 2010

NoC, and MPSoC devices) so to deal with power consumption and interconnect delay issues. The increased complexity impacts on design, development and management of new solutions, increasing both production costs and time to market. Therefore the creation and development of methodologies and techniques to support the design of complex systems and speed up related processes became a research field of main importance.

Hardware-Software Codesing is a well-established methodology in the Embedded System community when dealing with complex systems. In particular it allows distributing activities of the system between Hardware Components (various information processing devices), and Software Components (code executed on those processing units). This methodology presents a step-based design flow for the design and refinement of complex systems. The starting point is the behavioral specification, then a sequence of different optimizations and refinements is performed in order to achieve the best possible final implementation. Evaluations are carried out using functions of merit based on design criteria (e.g. performance, power, area or combinations of them), on solutions still fulfilling the main functional and non functional requirements.

On the other hand, the Model Driven Engineering (MDE) methodology, in which an overall system specification guides the developers along the design process, was traditionally used by the software engineering community. Appropriate standardized design languages (e.g. UML [22]), were used for this scope. MDE was recently extended also to the Embedded Systems field through appropriate UML profiles (as e.g. MARTE [18] or SysML [21]). Mechanisms are provided to describe structural, behavioral and non-functional specifications, in order to design a real-time embedded system from its model, supporting the entire development cycle.

MDE and Hw-Sw Codesign offer different complementary characteristics that in theory could solve many of the problems related to the increasing systems complexity. Unfortunately this complementarity is not fully exploited: very few tools (e.g. [20]) have the capability of joining both domains, and automated or semi-automated design flows are lacking. This work addresses such lack studying a practical semi-automated design flow in which both methodologies are merged and exploited to enable a fast design process. We aim at helping in designing Complex Real-Time Embedded Systems, executing several concurrent and sequential tasks. Our methodology described in the next sections involves Allocation and Binding within Design Space Exploration, Schedulability Analysis, Hw/Sw Partitioning, and Estimation techniques.

This chapter is organized as follows, Section 2 presents the state of the art and some related works. A general overview on the theoretical foundations of this project is given in Section 3. Section 4 describes our proposal of a semi-automated design work-flow. Section 5 analyzes the code generation process and the transformations defined to obtain SystemC simulators from MARTE models. Results obtained when this approach was applied to design a wireless voice link are shown in Section 6.

2. State-of-the-Art and Related Works

Model Driven Engineering [9] emerged as a methodology to manage the development of complex software systems. The origins of MDE could be traced down to the decade of 1980, when new design methods appeared to support the Object Oriented (OO) programming. The idea of describing systems by means of "particular domain" concepts (by an interactive *model*) instead of algorithmical specifications or computing concepts was rapidly accepted by the software community. The noticeable advantages of MDE have been continuously exploited leading to a mature design paradigm used worldwide, and promoting widely known approaches, like the Model-Driven Architecture (MDA) initiative by OMG [19].

The MDE methodology is achieved linking three main aspects, that represent the basic entities to support a development process. Namely *model*, *metamodel* and *model transformations* are the basis for any MDE activity, and they consist of:

- **Model** A model uses concepts and relations to reproduce an abstraction of the reality. Physical things become concepts in the model, and links of different nature between those things become relations. When the model is observed from different points of view specific aspects of the reality are highlighted. Another simpler definition of model can be found in [6]:

  > A model is a simplification of a system built with an intended goal in mind. The model should be able to answer questions in place of the actual system.

- **Metamodel** A metamodel offers a set of available concepts and relations to create a model, according to a particular purpose. A model is then said to conform to a metamodel. Metamodels defines the syntax of their models.

- **Model Transformations** A model transformation is the conversion of one model (source) into another model (target) using a set of rules. The rules map concepts and relations from the source metamodel to the target metamodel's concept and relations. A special case of model transformation is the code generation, that enables a model to be executable. This transformation is called *model-to-text* [2], and is based in a set of templates rather than in a set of rules.

Model nature is another key point of MDE. In particular, it allows to build different models of the same system, depending on how the system is looked at.

Referring to the MDA approach [31], we can find three default models of a system, as explained below:

- **Computation Independent Model (CIM)** The CIM focuses on the environment of the system, and the requirements for the system. The details of the structure and processing of the system are hidden or as yet undetermined. A CIM specifies the function (or external behavior) of a system without showing constructional details.
- **Platform Independent Model (PIM)** The PIM focuses on the operation of a system while hiding the details necessary for the implementation on a particular platform. A PIM shows the part of the specification that does not change from one platform to another.
- **Platform Specific Model (PSM)** The PSM combines the specifications in the PIM with the details that specify how the system uses a particular type of platform. These additional details are technical concepts, representing the different kinds of parts that make up a platform and the services provided by that platform.

By using CIM, PIM and PSM models it is possible to describe the system at different levels of abstraction. The implementation point is achieved applying transformations from model to model (CIM-PIM-PSM cycle) until sufficient details have been added. In the last ten years some studies tried to use MDE for Embedded Systems design. Real time analysis using UML-SPT profile [30] or other proprietary profiles, were carried out in [8, 12, 29, 32, 33]. More recent works use MARTE profile for such scopes. In particular Andre et al. [4] introduce the Time Model subprofile of MARTE; Taha et al. [35] introduce concisely the Hardware Resource Model profile and then presents an appropriate methodology to apply it during the hardware design process; Boulet et al. [7] highlight the expressiveness of MARTE notation for modeling regular distributions and clarifies its usage through examples and comparisons to other distribution notations such as in High Performance Fortran.

Hardware/Software Co-design

An embedded system is a compound of several hardware and software elements, where in general software is used for features and flexibility, while hardware is used for performance. Any design process willing to produce competitive results must consider both parts at the same time. Design activities have to take the existence of available platforms into account to adapt the behavioral characteristics and produce an optimal implementation. As a result a large number of activities become mandatory when designing an Embedded System [25], as described below:

- **Task Definition** Tasks representing the main processing activities in the system should be extracted and identified from the requirements. Such tasks or processes should be a representative set rather than the exact set used at implementation.

- **High-Level Transformations** Due to the heterogeneity of Embedded Systems it is mandatory to optimize high-level specifications through different transformations. Only in this way the additional benefits of some components can be exploited and improvements on performance, area, power or code size can be obtained. Some examples of these transformations are loop unrolling, loop tiling, array folding, and floating point arithmetic to fixed-point arithmetic convertion.

- **Hardware/Software Partitioning** Any task could be implemented using either software or hardware, but an analysis should be performed to find which kind of implementation is more favorable. By partitioning we understand the activity to map operations to hardware or software, obtaining the maximum benefit without exceeding the desired cost and constraints.

- **Time Ordering** The time ordering of tasks execution is another crucial activity when designing embedded systems. The *scheduling* assigns start times to operations, and allows to know at early stages if the available resources are enough to host the target tasks, and if they will lead to temporally correct executions.

- **Desired Solution Selection** The previously explained procedures usually yield several designs points or solutions satisfying the requirements, but as it can be anticipated not all these design points represent optimal implementations. The entire set containing all the possible implementations is known as the Design Space, and an exploration must be done to analyze it. One design point has to be selected among those satisfying the requirements, according to the designer experience. This procedure is known as *Design Space Exploration* (DSE).

Figure 2.1 presents what we can consider as a modern co-design flow, with a design loop covered many times before finding the desired implementation. This co-design flow is composed by elements and design steps that will be explained within the following subsections.

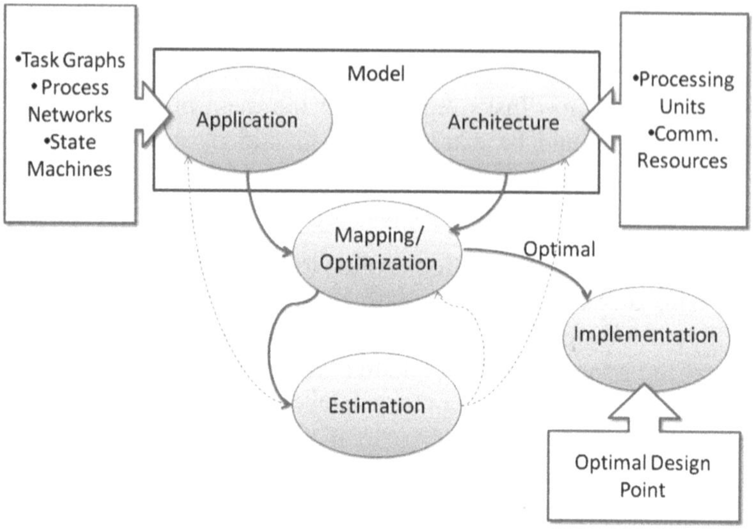

Figure 2.1. Hardware/software co-design suggested design flow

Related Work

Hw/Sw Co-design is a major focus in the literature and there were previous efforts to join such methodology with the techniques we use in this work. In particular in [1, 11] UML-based hardware/software partitioning is explored, while in [13] the scheduling analysis view is shown as an extension of the design to improve real-time predictability.

The most important contributions exposing complete design flows merging Hw/Sw Co-design and MDE can be found within the Gaspard2 toolchain in [24, 36], where the authors present a chain generating a SystemC simulation tool targeting High Performance Embedded Systems (HPES) modelled using MARTE profile plus some extensions. Another interesting work bridging Co-design and MDE is the theoretical proposal of MoPCoM [16], in [23], where a design process leads to SystemC code generation from MARTE model specifications. Although Gaspard2 and MoPCoM address some co-design problems, such as code generation and synthesis, they don't really provide support for the complete co-design cycle. In fact, those approaches leave outside essential concepts like Design Space Exploration, Allocation, Binding and Scheduling, that will be dealt with our approach.

In this work we started from our previous approaches to join MDE and Hw/Sw Co-design, and along with the expansion to support further Co-design techniques of the mathematical structure describing Design Space Exploration in SysML, exposed in [10, 26]; new SystemC code generation rules were created in addition to those shown in [27] aiming at mapping structural characteristics and other MARTE stereotypes. Finally our framework [28] was extended in order to allow better structural modelling and to add estimation techniques and scheduled execution to simulate real-time behavior.

3. Theoretical Context

Our previous works in Embedded Systems and SoC/MPSoC modeling, exposed in [26], were characterized by the usage of SysML as main design language, complemented with MARTE annotations for time and analysis modeling. This combination offers good flexibility when designing at high levels of abstraction (see e.g. [23]). In this work we chose to use only MARTE profile since it allows easier mapping for code generation, provides embedded systems specific model elements, and is fully supported by an open source tool, namely *Papyrus for UML*, which was used within this project. However, our semi-automated design flow can be used also with the SysML profile as the transformations described in [27] can be adapted to those shown in Section 6.

As mentioned before, the design process faces several challenges when dealing with modern hardware-software system. Application or target system complexity, and proliferation of numerous design objectives and constraints are the main issues that are traditionally dealt with Hw/Sw Co-design. A set of techniques are applied to a model in order to perform multiple optimizations and refinements, leading the system from an abstract stage to a physically implementable concept. Figure 2.1 represents the co-design flow and the components which are the basis of our MDE design process, as explained below:

- **Model** The system is represented by a model specifying both the *Application* and the *Architecture*. The applications is usually expressed in terms of Task Graphs (e.g. DFG, CFG), Process Networks (e.g. Kahn Process Networks, Synchronous Data Flow) or State Machines (e.g. StateCharts), depending on the models of computation. On the other hand the architecture specifies the processing units and the communication infrastructure of the system.

- **Mapping and Optimization** At system level, the mapping is composed of an *Allocation* (i.e. the selected processing units and communication resources), a *Binding* (i.e. assignment of tasks to selected components) and a *Scheduling* (i.e. execution order for the tasks). The allocation, binding and scheduling are produced on the basis of a multi-objective optimization criteria, examined through a Design Space Exploration (DSE). The process Mapping – Estimation – Optimization represents a recursive loop executed until the demanded cost-performance relation has been reached.

- **Estimation** After mapping, an estimation of the design properties of the next layers of abstraction should be made, so to anticipate the results due to design decisions at system-level. Accuracy of the estimation methods is a key point for a successful design.

- **Implementation** The output of the optimization process is composed of different design points, each one yielding cost-effective solutions (Pareto Points). The designer, based on the design objectives, should select one of these system configurations. The chosen system continues towards the development stage to be implemented.

4. Codesign-Driven Modeling: Integrated Design Flow

This section presents our design methodology, merging a system engineering process with Hw/Sw Co-design capabilities, as part of a semi-automated procedure. The implemented workflow, described in Fig. 2.2, features several model-to-model transformations that allow to make design decisions and enable simulation of the system behavior. The CIM-PIM-PSM cycle recommended by MDE is also followed through the entire process, making it possible to find a specific implementation starting from the system requirements. The flow starts with a *preparation phase*, which includes a requirements analysis, producing a set of functional and non functional specifications to be used in the model. If such information is not available it may be gathered through Rapid Prototyping, Simulation or Analysis of the System. In particular, the tasks Worst Case Execution Time (WCET) and/or Average Execution Time (AvgET) are mandatory for a proper design of a RTES targeting temporal correctness.

Requirements and non functional specifications, as identified in the previous step, are used as input for the *design phase*. The *design phase* consists of the realization of structural and behavioral diagrams (corresponding to Co-design Application and Architecture models) that expose all the possible *bindings*, using the MARTE Allocation Modeling to allocate in temporal or spatial fashion. Timing information is captured by means of qualitative and quantitative annotations contained in the SAM (Schedulability Analysis Model) package. In the next section we will define the design rules in order to further explain how the MARTE packages, diagrams, and stereotypes are used inside our design cycle.

The workflow continues in the *co-design analysis phase* with the generation of an XMI (XML Metadata Interchange) file from where data to perform a DSE on the allocation and binding options is gathered and provided to an Hw/Sw Co-design tool. All the possible bindings are first examined by a Schedulability Analysis tool in order to discard the combinations yielding time-incorrect (i.e. non respecting the real-time deadlines) solutions described within the requirements specifications. Then, among the remaining design points, after

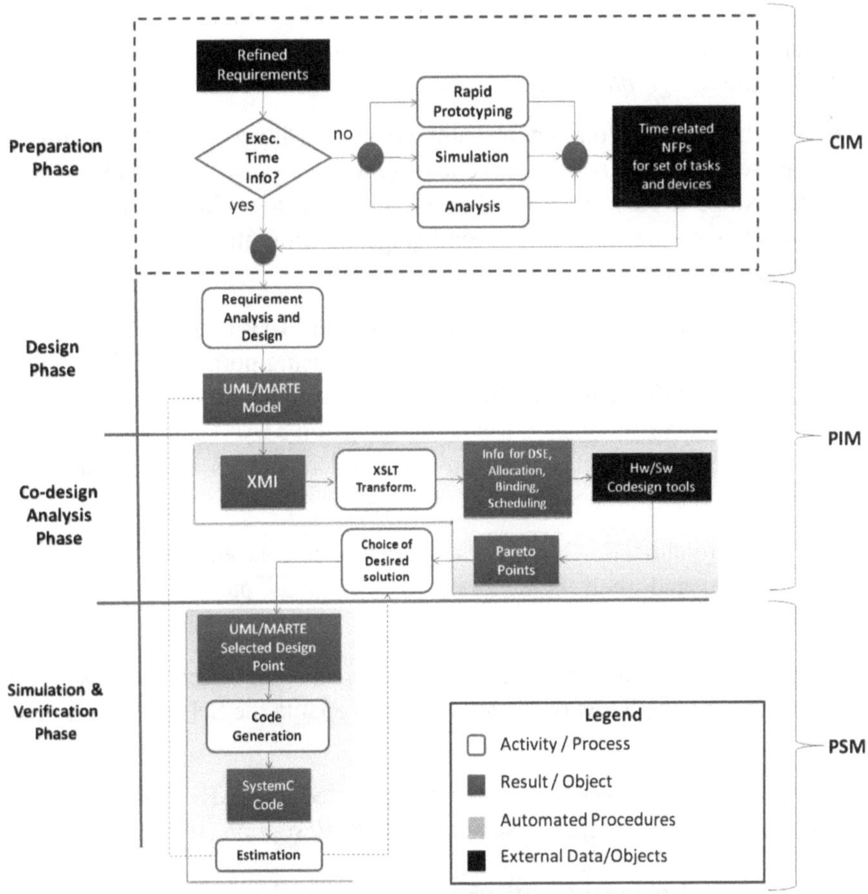

Figure 2.2. Semi-automated design work-flow

performing the multi-objective optimization through the DSE, the designer should select a Pareto-optimal point according to the desired cost criteria [5]. This design point, described by means of a UML/MARTE model is composed of an *allocation*, a *binding* and a *scheduling*. In the *Simulation* and *Verification* phase we generate the corresponding SystemC simulator necessary to check requirements, behavior and time correctness of the entire system.

Design Rules

Design rules are constraints yielding the selection of a suitable subset of MARTE to specify the system. The Embedded Systems domain is highly heterogeneous and it deals with several methods of computation, several application specific concepts and several architectural paradigms. The design rules, applied in the *design phase* of our workflow, are defined as follows:

- The behavioral components are modeled using both Task Chains expressed by *Activity Diagrams* and *StateCharts*. Activities are used in a first stage to identify tasks that are allocated to architectural resources. StateCharts are used in a second stage to refine the description of such activities completely defining the behavior.
- The structural components are defined with MARTE stereotyped classifiers and their communication is specified using ports and links, through a Composition Structure Diagram. The Architecture can be modeled at different levels of abstraction using the given MARTE subpackages, but the specific properties of highly detailed resources are not taken into account.
- The communication resources are described as any other processing resource, and their behavior is described as part of the functional specification.
- The architectural and behavioral components are stereotyped as recommended by the SAM package in order to enable the Schedulability Analysis of the system.

The next subsection shows how the rules converted to MARTE stereotypes are mapped to a mathematical representation of the *design phase*.

Mathematical Formalization and MARTE

A new mathematical structure of the model is formulated based on the theory exposed in [26]. Each term in the equations is mapped to a MARTE model element and/or a UML diagram. The formalisms describes the model M as a four tuple structure, as follows:

$$M = \{A, F, B, T\}$$

where:

1 $A = (V_A, E_A)$ is the architectural specification graph, composed of available hardware resources V_A and communication links between them $E_A \subseteq (V_A \times V_A)$.

2 $F = (V_F, E_F)$ is the functional specification graph, composed of tasks (procedures) V_F and edges $E_F \subseteq (V_F \times V_F)$ representing control flows.

3 $B = \{ (v_{Ai}, v_{Fj}) \subseteq (V_A \times V_F) \mid (v_{Fj}, v_{Fz}) \in E_F \Rightarrow (v_{Ai}, v_{Ak}) \in E_A$ $\forall (v_{Ak}, v_{Fz}) \in \mathrm{B} \}$ is a binding relation representing all the possible

Table 2.1. Correspondence between the mathematical formalism and MARTE elements

Notation	MARTE stereotype	UML element
V_A	*<<rtUnit>>* *<<HW_ComputingResource>>* *<<HW_Memory>>* *<<HW_StorageManager>>* *<<HW_Communication>>* and derived stereotypes	Classifier, object
E_A	*<<flowPort>>* *<<MessagePort>>*	Port
F	*<<RTBehavior>>*	Activity
V_F	*<<rtf>>*	Action
E_F	N/A	ControlFlow, transition
B	*<<allocate>>* *<<allocated>>*	Realization

bindings and allocations respecting the constraints imposed by control flows and communication links.

4 $T \subseteq$ (B $\times$ P) is the model elements characterization with non-functional properties (NFPs), where $P = \{p_1 \cdots p_n\}$ is a set containing NFPs as deadlines, execution times, priorities and scheduling methods.

This formalization and all its terms are directly mappable to diagrams and MARTE stereotypes. The main specification graphs A and F are described by means of *Composite Structure Diagrams* and *Activity/State Machine Diagrams* respectively. The components of the P set are defined as properties from MARTE::NFP subprofile. The NFPs are given by stereotypes from the Design Model package and the GAM/SAM subprofiles (e.g. *<<gaResourcesPlatform>>*, *<<gaWorkloadBehavior>>*, *<<gaWorkloadEvent>>* etc.). The rest of the structure is mapped to other UML components and MARTE stereotypes belonging to the Design Model, as shown in Table 2.1.

Co-design Analysis

A Hw/Sw Co-design Analysis is performed to select an optimal implementation after the *design phase* has been completed. Although this work is not for studying in depth co-design algorithms, it is worth giving a subtle remark on how those algorithms are reached by our UML/MARTE Model. This phase consists of a set of Model-to-Model transformations, performed using XSLT rules, so to create new representations that expose time related information. Hw/Sw Co-design tools for performing Schedulability Analysis and Design Space Exploration (i.e. SymTA/S [20], MAST [15], Cheddar [14], Multicube Explorer [17]) are fed with the resulting models in order to obtain Pareto Points yielding time-correct implementations. Following the design objectives an implementation candidate should be selected and simulated, to verify the behavior of the system. The automatic process generating the simulator is explained within the next section.

5. Code Generation: from MARTE Models to SystemC

The Automatic Code Generation process, explained in Fig. 2.3, is central for the Simulation and Verification phase. For such scopes we extended the model-to-text transformation capabilities of our tool SC2 [28]. A structural-oriented adaptation of the framework based on the Architectural Specification Graph (A) was included, in order to generate code for the SystemC macro modules

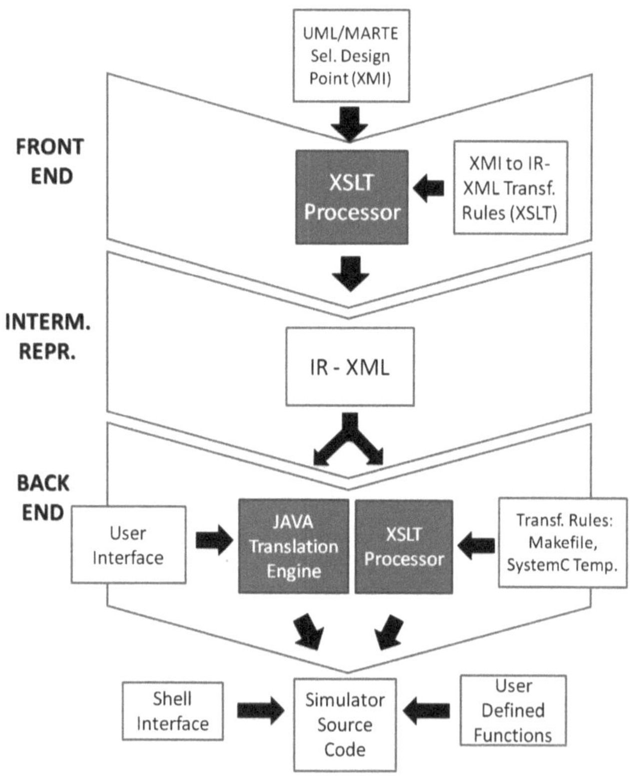

Figure 2.3. Automatic code generation procedure

composing the simulator (i.e. Hardware Resources). A scheduler triggering behaviors allocated to architectural resources is also included as a module connected to all Hw resources. The code generation from behavioral components is performed following our original philosophy. Some changes were introduced in the SystemC mapping rules and the simulator structure so to improve the communication between modules making it more flexible.

Architectural Extensions

The SC2 project consists of an automatic generator of simulators which generates RTL SystemC code from StateCharts based models. The layout of the generated code is close to a synthesizable application, and this fact can be exploited when performing several refinements to the model. The framework is composed of a compiler-like system, with Front-end, Intermediate Representation (IR-XML) and Back-end. XSLT transformations are performed in order to map the XMI file created by the UML tool into a standardized XML format (front-end) and then from such format to the identified SystemC template (back-end).

The produced simulators is made of modular SystemC code. The modules (i.e. SC_MODULEs in the code) are bound to different State Machines Diagrams in the model. In case a State Machines Diagram contains more FSMs (parallel components or hierarchical separation), each FSM is mapped to a SC_METHOD inside the same SC_MODULE. Injection of user defined functions, a shell interface for simulation, and multiple instance creation from a single model are supported by the tool. While such framework provides a good solution for mapping behavioral MARTE models, it lacks mapping rules for architectural components and structures. The extensions we present in this work are mainly targeted to support mapping from MARTE architectural information to SystemC code. We also added the annotation of NFPs inside the code for evaluation phase.

We modified the front-end so to allow recognition and parsing of MARTE stereotypes. This procedure involved also an improvement of the Intermediate Representation to incorporate such stereotypes (e.g. NFPs used inside the simulator) and collect when necessary the architectural information on the system. As we use Papyrus (because it fully supports MARTE profile), our front-end takes as input XMI files generated by this tool.

In the back-end more complex changes were needed to adapt the previous tool to our semi-automated flow. In particular modifications were necessary to redefine the structure of the generated simulator and support the inclusion of the hardware components as containers for behaviors (i.e. through <<allocate>> stereotype). The SC_MODULEs don't correspond to State Machines Diagrams as in previous version, but to MARTE hardware resources.

This change led to modify also the strategy used to communicate between State Machines. In the previous release it was not possible to customize inter-module communication. External modules were created corresponding to signals (declared in a class diagram); such modules were connected through ports to every module. In case of multiple instantiation the only possibility was to declare the variables as local or global. Local variables were connected to all the module of the particular instance, while global variables were connected to all the modules of all instances. In this work we enhanced such mechanism as we are able to represent the communication links between hardware resources. Thus the signals are connected to sc_port< > and therefore communicated to the appropriate hardware block when the architectural diagram dictates so. Internal variables are mapped to sc_signal< > and are only used inside the particular module.

Behavioral information contained in a State Machine diagram, mapped to SC_METHODs is used with the same transformation as in our previous tool. We added information coming from Activity Diagrams to control execution of the various tasks. A signal enables the scheduling of tasks. This signal is part of the sensitivity list of each SC_METHODs and triggers its execution. When the *task_id* identifier is received in the module (through a sc_port< > connected to the scheduler) the corresponding task is first notified and then unblocked and executed. The scheduler in turn analyzes which tasks should start at a given time (for each module in the system), and then triggers the execution. Finally, once the task has been executed, each module sends a message to the scheduler to communicate its availability.

If necessary task preemption can be implemented within this schema. Even though the SC_METHODs can't be stopped or restarted, our framework supports history. When a parent active state is created having all the state machines inside it and history enabled, the transition from this state to a suspend state emulates preemption. Using SC_THREADs would simplify the matter but it has a cost on simulation speed and performance as explained in [28].

Translation Rules

Mapping of MARTE stereotypes into SystemC code is essential to enable the translation process. The main rules are presented in Table 2.2. Such rules refer to immediate mapping in which it is possible to directly transform a stereotype into a SystemC artifact. Similar rules have been previously presented for UML and SysML transformations into SystemC by several studies (e.g. [3, 27, 34]).

6. Case Study

To show application of our complete methodology we take as a simple case study the design of a bidirectional wireless voice link with security provisions.

Table 2.2. Mapping of MARTE stereotypes in SystemC code

SystemC	MARTE stereotype
SC_MODULE	*<<rtUnit>>* *<<HW_ComputingResource>>* *<<HW_Memory>>* *<<HW_StorageManager>>* *<<HW_Communication>>* and derived stereotypes
sc_port< > sc_export< >	*<<flowPort>>* *<<MessagePort>>*
SC_METHOD	*<<rtf>>*
sc_signal< >	*ControlFlow, Transitions*
sc_clock	*<<clock>>*

We assume that we do not have the constraint that the two communicating nodes must be equal, therefore different bindings can be operated for them. The functionality of the system is as depicted in Fig. 2.4: a data packet is prepared, security operations are performed and then the packet is sent through a wireless channel. This packet is received and appropriately decrypted/authenticated, then it is decoded.

Our design space is described through the architectural graph drawn in Fig. 2.4. We can decide whether include a coprocessor for security or not (both in transmission and in reception) and whether to develop and use a dedicated radio protocol (i.e. RadioB), with a much higher cost, or use the already existing IEEE 802.15.4 (i.e. RadioA). The architectural graph dictates some constraints caught by the Binding relation, as an example it is not possible to choose the couple Radio1A and Radio2B.

The different choices have a different impact on the cost of the solutions as well as on the execution times. Data referring to WCETs are annotated in model through the stereotype <<GaScenario>> and are shown in the picture as comments for clarity. Therefore such data can be sent to a scheduler. We deal with a simple case but our approach supports also more complex situation with actions that are concurrently executed. In our case we have a deadline of 40 ms for the entire activity that we expressed through the <<saEndtoEndFlow>> and this together with the architectural constraints led us to isolate valid bindings. The solution we took was to have 802.15.4 radios and security coprocessor mounted only in the first device.

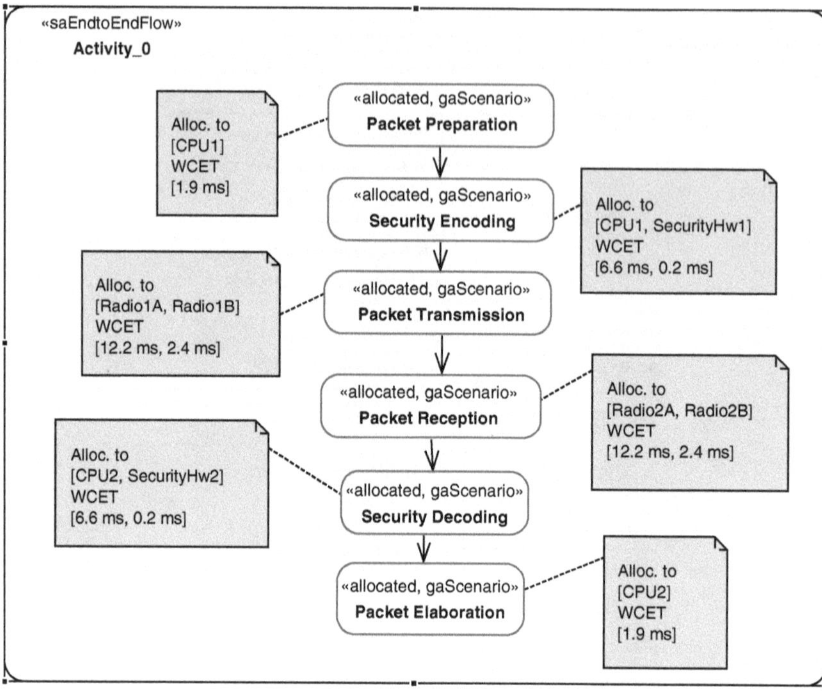

Figure 2.4. Functional specification graph of our application and architectural graph representation of our design space

Having decided our solution the next step consists in refining the behavioral description completing the State Machine (StateCharts) diagram corresponding to the desired configuration. The framework provided us with a working simulator of our solution. We could use it to try with a characteristic data stream that the solution we chose respected the design constraints.

7. Conclusions and Future Works

In this chapter we propose a framework that extends our previous works and brings them together in a wider research context in order to join the benefits of Hw/Sw Co-design and MDE. We show how it is possible to compress in a few diagrams complete information about the design space, feeding through convenient stereotypes a Hw/Sw Co-design tool to calculate Pareto Points. Moreover behavioral description through StateCharts is used to automatically generate simulators that are useful for evaluating systems performance.

Using a convenient subset of MARTE UML profile we support the design flow from Design Space Exploration and Scheduling until automatic

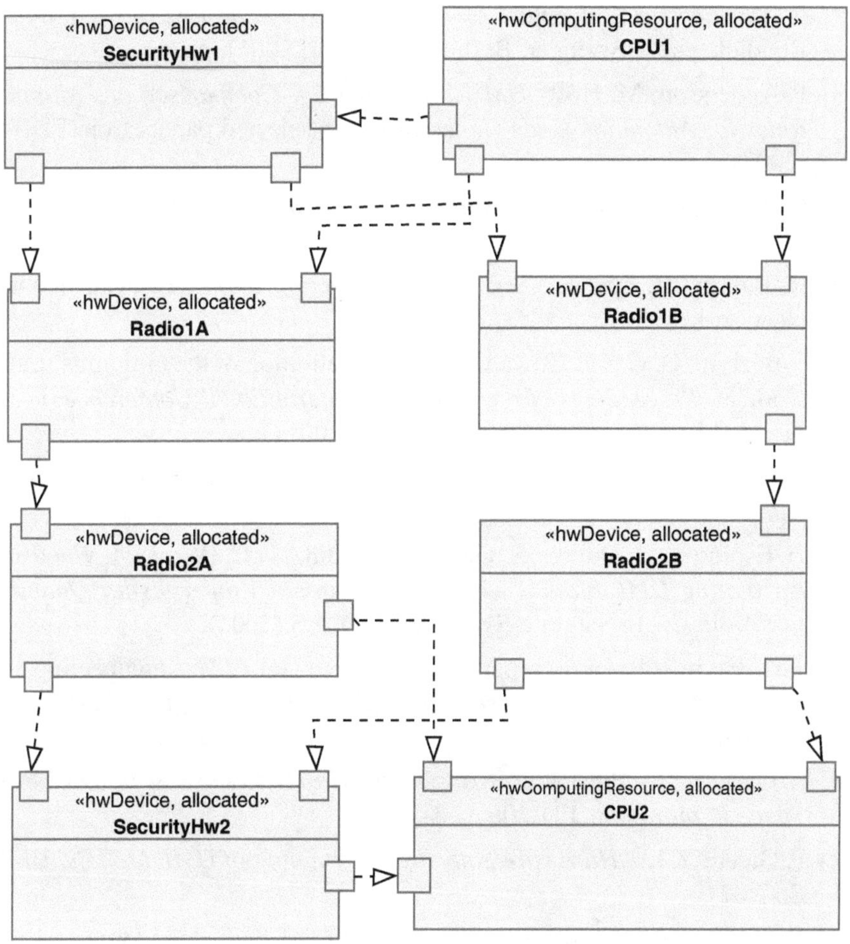

Figure 2.4. (continued)

generation of SystemC Executable Models. Future work involves extension of the supported subset of MARTE profile and its mapping to more SystemC constructs as e.g. sc_fifo< >.

References

[1] W. Ahmed, D. Myers, *Using UML 2.0 for the Creation of Concept-Based SoC Design.* UML-SoC'07

[2] M. Albert, J. Munoz, V. Pelechano, O. Pastor, Model to text transformation in practice: generating code from rich associations specifications,

in *Advances in Conceptual Modeling – Theory and Practice*, ed. by J.F. Roddick, et al. (Springer, Berlin/Heidelberg, 2006), pp. 63–72

[3] P. Andersson, M. Höst, *UML and SystemC – Comparison and Mapping Rules for Automatic Code Generation.* On selected papers from FDL'07 (2007)

[4] C. Andre, F. Mallet, R. de Simone, *Time Modeling in MARTE.* FDL'07 (2007)

[5] R. Bayton, R. Spence, *Sensitivity and Optimization* (Elsevier Science, New York, 1980)

[6] J. Bézivin, O. Gerbé, Towards a precise definition of the omg/mda framework, in *Proceedings of the 16th IEEE International Conference on Automated Software Engineering, ASE'01* (2001)

[7] P. Boulet, P. Marquet, E. Piel, J. Taillard, *Repetitive Allocation Modeling with MARTE.* FDL'07

[8] H. Espinoza, H. Dubois, S. Gerard, J. Medina, D.C. Petriu, M. Woodside, *Annotating UML Models with Non-Functional Properties for Quantitative Analysis.* In Satellite Events at MoDELS (2005)

[9] J.M. Favre, Towards a basic theory to model driven engineering, in *Proceedings of Third Workshop in Software Model Engineering, Lisbon, Portugal* (2004)

[10] S. Ganesan, M. Prevostini, *Bridging the Gap Between SysML and Design Space Exploration.* FDL'06

[11] P. Green, Y. Lu, *Hardware/Software Partitioning of UML Models.* UML-SoC'07

[12] P.N. Green, M.D. Edwards, *The Modeling of Embedded Systems Using HASoC.* In Proc. of DATE 02

[13] M. Hagner, M. Huhn, *Tool Support for a Scheduling Analysis View.* UML Workshop at Date'08

[14] http://beru.univ-brest.fr/ singhoff/cheddar

[15] http://mast.unican.es

[16] http://www.mopcom.fr

[17] http://www.multicube.eu

[18] http://www.omgmarte.org

[19] http://www.omg.org/mda

[20] http://www.symtavision.com

[21] http://www.sysml.org

[22] http://www.uml.org

[23] A. Koudri, D. Aulagnier, D. Vojtisek, P. Soulard, C. Moy, J. Champeau, J. Vidal, J.C. Le Lann, *Using MARTE in a Co-Design Methodology.* UML Workshop at Date'08

[24] P. Marquet, S. Meftali, S. Niar, A. Etien, J. Dekeyser, E. Piel, R. Attitalah, P. Boulet, *Gaspard2: from MARTE to SystemC Simulation.* proc. DATE 08

[25] P. Marwedel, *Embedded System Design* (Springer-Verlag, New York, 2005)

[26] M. Mura, L.G. Murillo, M. Prevostini, *Model-Based Design Space Exploration for RTES with SysML and MARTE.* FDL'08

[27] M. Mura, A. Panda, M. Prevostini, *Executable Models and Verification from MARTE and SysML: a Comparative Study of Code Generation Capabilities.* MARTE UML Workshop at DATE'08

[28] M. Mura, M. Paolieri, *SC 2 StateCharts to SystemC: Automatic Executable Models Generation.* FDL'07

[29] M. Natale, M. Saksena, *Schedulability Analysis with UML,* in *UML for Real: Design of Embedded Real-Time Systems* (Kluwer, Norwel MA, 2003)

[30] Object Management Group, *UML Profile for Schedulability, Performance, and Time.* In OMG document n. ptc/02-03-02 (2002)

[31] Object Management Group, *MDA Guide, Version 1.0.1.* OMG document number omg/2003-06-01 (2003)

[32] D.C. Petriu, C.M. Woodside, Performance analysis with UML: layered queueing models from the performance profile, in *UML for Real: Design of Embedded Real-Time Systems* (Kluwer, Norwell MA, 2003)

[33] T.H. Phan, S. Gerard, D. Lugato, *Scheduling Validation for UML-modeled Real-Time Systems.* ERCT (2003)

[34] W. Raslan, A. Sameh, *Mapping SysML to SystemC.* FDL'07 (2007)

[35] S. Taha, A. Radermacher, S. Gerard, J.L. Dekeyser, *MARTE: UML-Based Hardware Design from Modeling to Simulation.* FDL'07

[36] R. Atitallah, E. Piel, J. Taillard, S. Niar, J.L. Dekeyser, *From High Level MPSoC description to SystemC Code Generation*, ModEasy'07 (2007)

Part II

C/C++-BASED SYSTEM DESIGN

Part II of this book addresses language-based modelling and design techniques for simulation, debugging, transformation, and analysis of hardware/software embedded systems. C/C++ based design methodologies are entering productive industrial design flows especially after the IEEE standardization of SystemC. Hence, the lion's share of contributions uses SystemC and its extensions to illustrate the scientific approach. However, articles using languages like graphical (UML) or functional languages (e.g. Haskell) were presented, covering interoperability issues and heterogeneous models of computations. Other topics were embedded software modelling or technology specific approaches, e.g., for reconfigurable computing or signal processing platforms. Efficient mechanisms for abstraction and simulation like transaction level modelling (TLM), IP-based system design or system synthesis were also covered by this workshop.

In Part II, you will find six excellent articles representing the typical spectrum of contributions to this thematic area. Hopefully, the following overview helps you to pick your favourite articles.

In CHECKPOINT AND RESTORE FOR SYSTEMC MODELS, Monton et al. present an efficient technique called checkpointing to save and restore the state of a SystemC simulation model. Rather than saving the complete model state, their approach aims at saving only essential information needed to restore the state. Models that follow some coding guidelines can reliably be stored and restored with checkpoints that are orders of magnitude smaller than the full model.

The exploration of design alternatives for multi-processor SoCs was one of the main motivations for SystemC TLM based modelling. In their paper EFFICIENT APPROXIMATELY-TIMED PERFORMANCE MODELING FOR ARCHITECTURAL EXPLORATION OF MPSOCS, M. Streubühr et al. present a framework for modelling MPSoCs at Electronic System Level (ESL) where the mapping of the application and the platform model are configured at elaboration time. The benefits for faster design space exploration are shown in an example.

The memory subsystem has a huge impact on the performance of many embedded systems. The efficient analysis of embedded memories is challenging as it requires fast and correct models. Loeb and Sauer present in their article FAST SYSTEMC PERFORMANCE MODELS FOR THE EXPLORATION

OF EMBEDDED MEMORIES an extension to the SystemC-based SystemClick framework. It allows combining different memory models to accurately reflect the performance characteristics of embedded memory. In an MPSoC example they evaluate the technique and compare it with state-of-the-art techniques.

Most embedded systems are compositions of heterogeneous components which often require combining different models of computation to efficiently describe such complex systems. The article ANOTHER TAKE ON FUNCTIONAL SYSTEM LEVEL DESIGN AND MODELING from Toczek, Houzet, and Mancini proposes a Haskell-based approach relying on functional programming principles to face the challenges when designing very big and complex systems. In an application example they validate the applicability of their modelling technique.

Reconfigurable technologies are highly flexible but require new design methodologies as standard techniques cannot cope adequately with adaptivity. In GENERIC MODEL FOR APPLICATION-SPECIFIC PROCESSORS ON RECONFIGURABLE FABRIC, A. Yurdakul el al. present a framework containing a new model, a design language (LRH+) and a set of tools for the efficient automated design of runtime-reconfigurable processors. Based on an application model in LRH+, the presented framework allows for generating a specific instruction set to be implemented on a runtime-reconfigurable fabric, exploiting the full flexibility of this technology.

Most current C-based synthesis tools restrict the designer either to the OSCI SystemC Synthesis subset or to a sequential ANSI C/C++ model. In A SYSTEMC SUPERSET FOR HIGH-LEVEL SYNTHESIS, M. Smirnov and A. Takach propose to combine these modelling styles into a new concept. Their approach not only addresses high-level synthesis but also stepwise refinement and enables simulation at different levels of abstraction.

Frank Oppenheimer
FDL'09 CSD Technical Area Chair
OFFIS, Oldenburg, Germany

Chapter 3

CHECKPOINT AND RESTORE FOR SYSTEMC MODELS

Màrius Monton[1,2], Jakob Engblom[3], Christian Schröder[4], Jordi Carrabina[1], and Mark Burton[2]

[1] *CEPHIS, Universitat Autonoma de Barcelona, Barcelona, Spain;*
marius.monton@uab.cat; jordi.carrabina@uab.cat

[2] *GreenSocs Ltd., London, UK;*
mark.burton@greensocs.com

[3] *Virtutech Inc., Stockholm, Sweden;*
jakob@virtutech.com

[4] *Department E.I.S., Technical University of Braunschweig, Braunschweig, Germany;*
christian.schroeder@greensocs.com

Abstract We present preliminary work in the field of saving and restoring model state within a SystemC simulation environment. Save and Restore (or checkpointing) is a useful technique that can greatly assist target software and simulation model development and debug. In contrast to other approaches that aim at saving and restoring the state of an entire simulation process, we investigate mechanisms by which only the essential model state is saved. This makes the checkpoints far more compact, saved simulation states can be moved between host machines, and be used with updated simulation models. Our initial results indicate that SystemC models written to certain coding guidelines can be saved and restored reliably. As a result, virtual platforms and platform components written in SystemC can be made more useful to software developers, and support better workflows.

Keywords: SystemC, TLM-2.0, virtual platform, checkpointing

D. Borrione (ed.), *Advances in Design Methods from Modeling Languages for Embedded Systems and SoC's*, Lecture Notes in Electrical Engineering 63,
DOI 10.1007/978-90-481-9304-2_3, © Springer Science+Business Media B.V. 2010

1. Introduction

Checkpoint save and restore (usually known as "checkpointing") is a process by which a simulator stores the state of the simulated system to disk, and later loads it back into the simulator, resulting in the exact same simulated system state. Checkpointing is also used in some other software systems, such as databases or long-running computational programs, to allow a run to be suspended or recovered after an error. In the case of virtual platforms, checkpoints have to include the contents of memories and disks, the state of processors, peripheral devices, and network connections in the virtual system, as well as the state of the simulation kernel including current time and any event queues and simulation scheduler state.

Checkpointing is a key workflow enabler for software development on a virtual platform. With checkpointing, a software developer can save and restore their work at any point and resume it later without having to keep the simulator running. A repetitive procedure such as booting an operating system and loading a set of software applications onto the simulated system can be done once (sometimes known as a "nightly boot"). Subsequently the configured system can be immediately resumed from that state, saving time and ensuring multiple developers have the same system. This is getting more and more important as simulated systems increase in complexity and workload size. We have seen cases where a system bring-up takes hours of real-world time, as it involves the simulation of tens of billions of instructions across hundreds of processors. Needless to say, in these cases checkpointing is necessary to avoid spending all working time booting the target system [1]. Checkpointing also makes it possible to store a library of booted and configured systems of various forms, for use in regression testing or to try various alternative micro-architectures on the same booted software load.

A checkpoint has to be portable across hosts and simulator versions. As checkpoints are exchanged between different user groups and different companies, it is impossible to know where they might end up. As an example, it must be possible to use the same checkpoint on a 32-bit little-endian x86 Linux host and on a 64-bit big-endian UltraSPARC Solaris host. It is also very useful if it is possible open a checkpoint taken in an earlier version of a virtual platform in a later version.

During simulator development, checkpoints make it easy for customers of a modeling service to report bugs and test the fixes provided. By using a checkpoint of a situation where the model fails to behave correctly, it is very easy for the modeling service to reproduce the issue. Once the model has been updated, the same checkpoint can be used to ascertain that the target system progresses beyond the point where it previously exhibited incorrect behavior.

In a similar vein, checkpoints are useful during iterative model development where most parts of a model are marked as "unimplemented". As soon as software actually accesses unimplemented features, a checkpoint is taken, the model is updated and simulation restarted from the checkpoint to see that the software reacts correctly to the now implemented register or feature. This enables very rapid development of virtual platforms, especially for existing hardware systems [1, 4].

Checkpointing can also be used to change the level of abstraction of a simulation model. A target system is booted and configured using a fast virtual platform, after which a checkpoint is taken. That checkpoint is then used to initialize the state of a detailed (typically, cycle-accurate) model of the same system. From a single checkpoint, several differently configured detailed simulations can be started, allowing efficient parallel exploration of the architecture space. Compared to switching a model between abstraction levels during a simulation run, this offers a more reliable, efficient and precise mechanism that also preserves the initial state of the detailed simulation for future regression testing. By taking many checkpoints during a fast simulation run, it is possible to sample the execution of a large system [7, 14].

Checkpointing Implementation Issues

To support checkpointing in SystemC (and indeed in any simulator system), there are three problems that have to be solved:

1 Saving and restoring the simulation state of all models in a simulation. Model properties like register contents, current state of state machines, and similar, must all be saved.

2 Saving and restoring the simulation kernel state, such as event queues and the current simulation time.

3 Saving and restoring the simulation configuration in terms of which simulation models form part of a virtual platform, and how they are connected.

We have addressed these problems for hardware device models written in SystemC™, using the OSCI reference SystemC simulator version 2.2.0. To create a complete system involving processors, memories, and software, we used Virtutech Simics [10] to provide us with the rest of the platform, as well as a proven checkpointing infrastructure. That let us focus on the core technology issues in SystemC model checkpointing rather than checkpoint file structure and disk I/O.

2. Simics Checkpointing Basics

To understand our implementation, some background on Simics is needed. Virtutech Simics has implemented checkpointing for about ten years, based on a simple but powerful mechanism called attributes [4].

Each simulation model in Simics defines its own set of attributes, which are expected to define the entire model state needed to continue the simulation from a particular point (Fig. 3.1).

The attributes are registered with the Simics kernel by the model source code at simulation start-up, and have a name and a type. The attributes are set and retrieved from the simulation kernel using an interface distinct from the transaction-level interfaces used for memory accesses and device-device communication. Figure 3.2 shows a simplified view of device models and connections between them in Simics.

Attributes do not have a one-to-one correspondence with variables and data structures used in the implementation of the model. Internal caches and data structures are often used for efficient simulation, but they would be represented as simple values or lists of values in a checkpoint. An important point to note is that since this decouples the externally visible model state from the implementation, different implementations of a model can load the same set of attribute values. This makes checkpoints independent of model implementations, and it is quite common for model implementations to change while attribute sets and checkpoints remain constant. It also enables the use of the same checkpoints

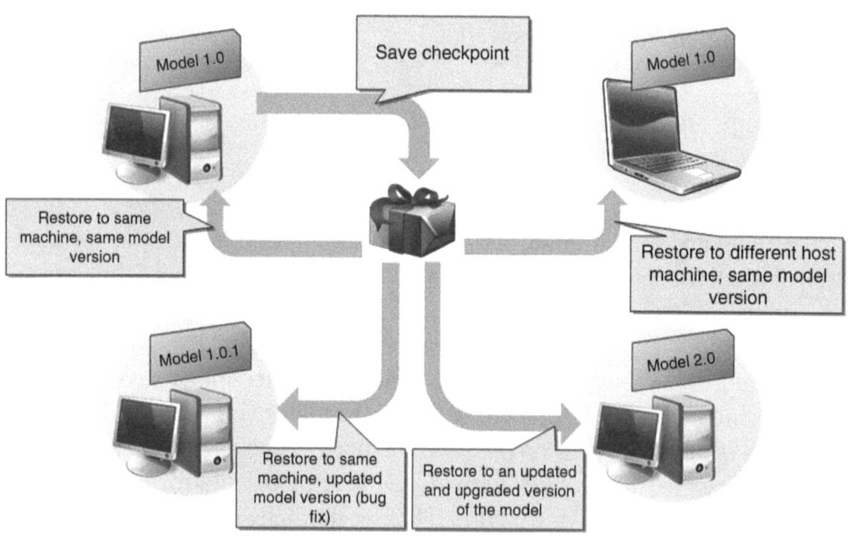

Figure 3.1. Checkpointing use cases

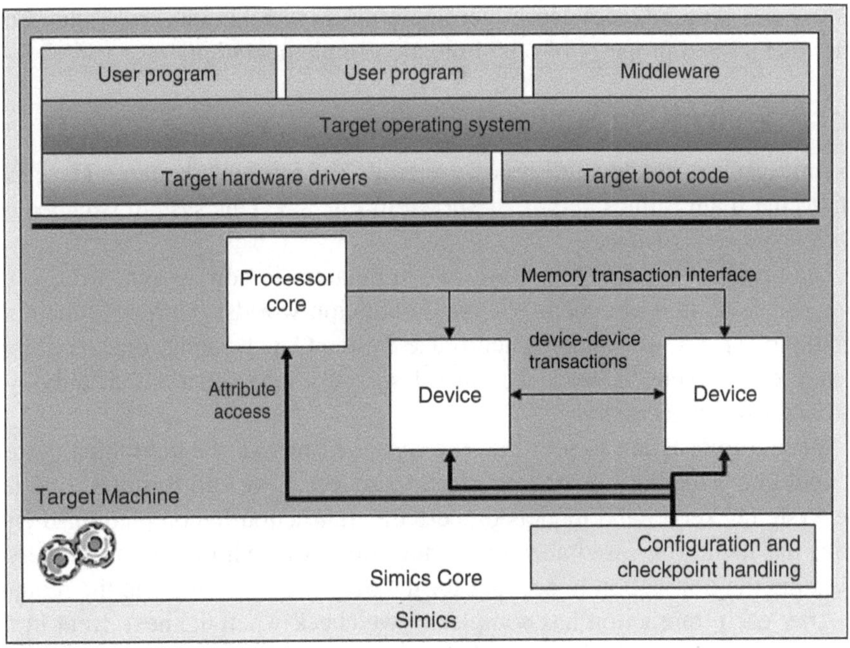

Figure 3.2. Simics back-door attribute system

for different levels of abstraction of the same model. Typically, a checkpoint represents the architectural state of a model, not its implementation.

Simics also represents the configuration of a system by means of attributes, converting pointers between simulation models into names, and storing names and types of model instances in checkpoints.

By using the Simics platform, we have been able to focus on the particulars of how to save and restore the state of SystemC device models, and not deal with how to put information in a file or the complexities of checkpointing the state of a processor model (as we use Simics existing processors and system models as the framework to run target software interacting with our SystemC models).

When an attribute is accessed in Simics, the simulation module gets a call to a special attribute getter and setter functions. We used this mechanism to export the state of our SystemC models to Simics, and later push the saved state back into the SystemC subsystem. Thus, all our SystemC models need to do is provide a means to expose the state of models and the SystemC kernel to Simics.

Initially, we have considered the entire SystemC simulation setup as a single Simics "model", with a fixed internal configuration. Thus, we do not attempt to convey the SystemC simulation configuration in checkpoints, but rather consider that a fixed property of our simulation implementation.

3. SystemC in Simics

To run SystemC models in Simics, we made the SystemC kernel a slave system to the main Simics kernel, as shown in Fig. 3.3. This system is called the *Simics-SystemC Bridge.*

The key problem in integrating two simulators is how to synchronize the time and execution of code in the two simulation worlds. In order to maintain simulation speed, we cannot synchronize SystemC with Simics on every target system clock cycle. Instead, we have designed a lazy scheme that only synchronizes when necessary.

When a transaction is sent into the SystemC part of the combined system (it could be a memory operation, reset, input, etc.), we call SystemC and lets the SystemC kernel and models process the transaction to completion. If SystemC has then progressed ahead of Simics time, we stall the Simics processor initiating the transaction to account for the time taken to complete the request.

After each transaction has completed, we check when the next event in the SystemC kernel is scheduled. If there is such an event, we post an event in the Simics event queue at a corresponding point in time. When the Simics

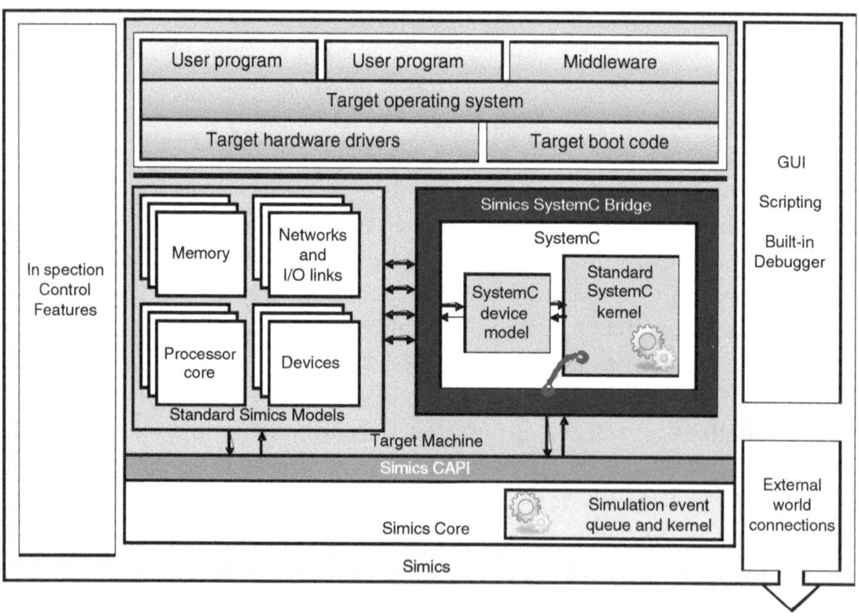

Figure 3.3. SystemC Encapsulated within simics

event is triggered, the SystemC kernel is invoked so that time can catch up with Simics time and the event be processed. Thus, SystemC models can perform actions asynchronous to the Simics world, like serial console output or sending completion interrupts [5].

The Simics-SystemC Bridge also has to convert transactions between their Simics and SystemC types. This is a reusable feature for simple standard interfaces such as TLM2.0 [11] or GreenBus [9], but for other interfaces you need a case-by-case conversion. The performance of the integration is very good, as SystemC does not impose any slowdown on Simics when SystemC device models are not being activated. Also, the overhead of translating transactions is minor compared to the cost of actually computing the result of a transaction in a device model.

To handle checkpointing, the Simics-SystemC Bridge is extended to get callbacks upon checkpoint save and restore in Simics. It also needs to convert from the state representation in SystemC to Simics attributes, and register attributes with Simics.

4. Checkpointing SystemC Model State

Model Requirements

Checkpointing requires a model to explicitly define the state that will be saved and restored. In our implementation, we have used the GreenControl [15] parameter mechanism to simplify the implementation in SystemC and make the changes to the SystemC code as small as possible.

GreenControl lets us mark variables in a SystemC module as managed parameters using the `gs_param<>` template. From SystemC code point of view, such variables can be used just like any other variable, with the added benefit that its value can be accessed and changed from the GreenControl system. The GreenControl system allows access to parameters from outside the module that declares them, and even when the simulation is not running. Figure 3.4 shows a short example of the use of parameters. Essentially, this provides the backdoor access we need to get and set the simulation state.

When a `gs_param<>` is declared, it is given a parameter name, which is used to access it along with an automatic hierarchical name derived from the SystemC system setup.

SystemC Parameters to Simics Attributes

We implemented and tested two ways to connect the parameter handling in GreenConfig to the Simics attribute system.

Variant one is to create a Simics attribute for each parameter in the SystemC device. This makes the checkpoint very readable, and has the side benefit of

```
class example : public sc_module {
public:
SC_CTOR( example ) :
  scparam ("scparam", 0xdeadbeef)
  {
    // ...
  }
//...
private:
  gs_param<uint32_t> scparam;
};
```

Figure 3.4. Example use of gs_param<>

```
simics> ckpt0->gs_all_param_value
"systemc_greencheckpoint_test.otherparm=42;
systemc_greencheckpoint_test.scparam=66;
systemc_greencheckpoint_test.scparam_two=4711;"
simics> ckpt0->scparam
66
simics> ckpt0->scparam_two
4711
```

Figure 3.5. Simics CLI session with the two styles of attributes

exposing the state of the SystemC devices for easy modification and inspection from the Simics user interface. It suffers from the need to add code to define one attribute per parameter, however.

Variant two is to use the GreenControl feature of exporting all parameters in a simulation as a list, and then convert that to a list attribute in Simics. This solution makes it somewhat more cumbersome to access the state from the Simics user interface as you have to parse a list of key-value pairs, but it also means that a single attribute is all that needs to be defined, which makes the solution independent of the SystemC model internals. Both solutions work equally well for checkpointing. The user-perceived difference is shown in Fig. 3.5. We have used the list variant, as it is more flexible.

Limitations

The main limitation to the system proposed above is that only SystemC module data members using the gs_param<> mechanism are checkpointed. In particular, signals and ports are not currently covered by our implementation. This limitation means that signals and port values are not saved, and SystemC modules have to be written with this in mind. For modules written in a TLM style, this is not a big issue since communication between modules is based on function calls that finish each transaction as a unit and that naturally store the updated module state in parameters marked with `gs_param<>`.

Another limitation is that we currently assume the SystemC setup to be a fixed subsystem in the Simics system. The entire SystemC subsystem is represented as a single Simics simulation model, with attributes reflecting the state of the SystemC kernel as well as the SystemC device models and the SystemC bridge itself. You cannot change the structure of the SystemC subsystem and expect checkpoints to work, as that will likely change the names of the parameters. This limitation could be addressed by splitting the SystemC subsystem apart into a set of Simics modules.

5. Checkpointing SystemC Kernel State

SystemC is an event-based simulator that maintains the simulation state storing pending events for `SC_METHOD`s and `SC_THREAD`s in separate priority lists. When simulation begins, the kernel finds the top event of each list and executes or resumes the process sensible to that event. We need to save and restore all information that the SystemC kernel needs to run a simulation. For this, we need to save the current simulation time and the event queues. Currently, this works only for `SC_METHOD`s, because we cannot access all the state of `SC_THREAD`s.

In case of `SC_METHOD`s, the SystemC runtime model is to call the sensitive "process" when an event to which the module is sensitive occures. Essentially, this becomes an event-fired function call into the device model, is executed. Once the function called returns and the execution of the sensitive process ends, the `SC_METHOD` module is suspended until another event that it is sensitive to occurs.

A more complicated mechanism is used when managing `SC_THREAD`s because a process can be suspended in middle of its execution (when `wait()` is called). This requires the use of a user-level threading system to manage the execution of the model. Execution of the thread is later resumed at the point where it was suspended, when an appropriate wakeup event happens. The problem for checkpointing of `SC_THREAD`s is not really the SystemC kernel state related to the threads, but rather the threads and their stacks in themselves. It is infeasible to access, not to mention restore, the stack-based state of an `SC_THREAD`.

Fundamentally, the use of suspendable threads is inappropriate for checkpoint and restore. It puts state in stack-allocated local variables, as an implicit part of the call stack, and into processor registers. This means that state is not explicit and not available for manipulation from the outside. Also, the layout of stack frames and local variables depends on the particular compilation of a particular version of the code, and is likely to change arbitrarily from even very small code changes. It ties the state of the simulation to the implementation of the simulation model in such a way that they cannot be separated.

The resume operation in our prototype consists of constructing the SystemC subsystem again (as it would be starting a new simulation), and once the elaboration is done, we update the kernel state to the state saved in the checkpoint. In particular, the simulation time is changed, and pending events are reposted to the event queues. Thus, the SystemC kernel will run the simulation just as if it were just continuing the checkpointed simulation without any interruption. Thanks to the event-driven style of `SC_METHOD`s, this works very well.

We access the event queue through the OSCI SystemC kernel class `sc_simcontext`. This class encapsulates SystemC kernel and it is in charge of managing the simulation, and provides us with the time and handle to the event queues. We also had to add a method `get_elem(i)` to the SystemC utility classes `sc_ppq_base` and `sc_ppq` to non-destructively read out the list of pending events. The existing `insert()` method was used to insert the saved list of events into the event queue.

The list of events from the SystemC kernel is communicated to Simics as a Simics attribute of the SystemC bridge module (Simics supports arbitrarily nested lists as an attribute type, which makes it easy to represent the event queue).

When a checkpoint is taken in Simics, the event queue attribute is read and stored using the Simics checkpointing mechanism. When resuming, the Simics attribute is written with the value (old event queue state) stored in the Simics checkpoint file, and our SystemC bridge extensions then repost the events into the new SystemC simulation, recreating the state at the point when the checkpoint was taken.

6. Related Work

The use of checkpointing for switching between simulation abstraction levels was first explored in the SimOS simulation system in the mid-1990s [14]. Today, in addition to Simics, the IBM Mambo full-system simulator [12] uses a checkpointing mechanism similar to that in this paper. Typical uses for checkpoints in Mambo are to save complex system setups and to switch between different implementations and different abstraction levels. Simics checkpoints have also been used to implement abstraction-level switching and sampled simulation [7].

The Boost C++ serialization library can save and restore the state of C++ objects in a program [13]. It provides portability and upgradability to new versions of the code. The Boost library only saves and restores data stored in C++ objects allocated on the heap, not temporary values on the stack, nor the state of threads in the system. Using this library with SystemC would have required a rewrite of the SystemC kernel to use serialized objects to store all state, and it is not clear how it could be extended to handle the threading in the SystemC kernel.

Another alternative solution that has been proposed is to save the entire contents of the memory of a running simulation process. At least in theory, this should work for any code, without modification, and including thread state. This solution is apparently used by CoWare in their announced checkpoint support for SystemC, but no details to verify this have been published [3]. Cadence has also introduced a save/restore mechanism for SystemC code in their Cadence Incisive Simulator [2]. The Berkeley Lab Checkpoint/Restore system for use in Linux-based clustered high-performance computing is a more powerful solution that can save and restore entire sets of threads, based on using a Linux kernel module [8].

However, the process-dump solution was actually tried in early version of Simics, and found to be very brittle and limited. The state stored depends on the details of the compiled program, as offsets on the stack and the order of variables in memory can vary from compilation to compilation. Such solutions also require all dynamically linked libraries to be identical, as well as the operating system kernel. In practice, it only works to restore onto the same machine, in the same implementation of the model, and thus does not fulfil the portability and upgradability requirements.

An even more heavyweight solution is to place the simulation inside a VMWare virtual machine, and use the whole-machine snapshot function of VMWare to save and restore the simulation. This works, but you cannot change the code, and the size of the snapshot is the size of the total memory of the virtual machine, which is usually on the order of gigabytes.

7. Experiments

We have performed a range of experiments to measure the overhead of our solution, and to test that it can achieve what we expect it to.

Simics-SystemC Bridge Performance

The performance of the basic bridge itself was tested by integrating an NS16550 serial port modelled in SystemC in a Simics virtual platform for the PowerPC 440GP SoC (an AMCC "ebony" board). The ebony board runs a full Linux kernel and U-Boot. One of the serial ports on the board was replaced with the SystemC serial port. Tthe performance impact of using the SystemC model compared to the Simics standard NS16550 model was imperceptible in normal use. When using a microbenchmark program that pushed characters onto the serial port as quickly as possible, the performance was reduced by about five percent. Thus, we conclude that the SystemC Bridge does not present a significant performance problem for typical TLM models.

Table 3.1. Performance test result for gs_param

Data type	Runtime (s)	Relative runtime
unsigned int	15.48	1.00
gs_param<unsigned int>	15.76	1.02
sc_uint	15.79	1.02
gs_param<sc_uint>	16.18	1.05

SystemC Checkpoint Support Overhead

In our SystemC implementation of checkpointing, there is a potential additional performance impact from the use of `gs_param<>`, since using the GreenControl mechanisms.

Because we assume frequent accesses to the variables marked as checkpointable during simulation runtime, the usage should be efficient. Neither the existence of checkpoint parameters nor frequent read or write accesses (when running the simulation, not when taking a checkpoint) should slow down the simulation significantly.

We created a performance test [15] (cf. tbl. 1) consisting of two TLM-2 devices: A master exercises 100 million TLM-2 write transactions to a slave which writes the received payload to an unsigned int, a `sc_uint` or `gs_param` variables. This means an intensive usage of parameters inside the tested module.

Table 3.1 shows that accessing model checkpoint variables which have been instrumented with `gs_param` to be quite efficient. The results show a penalty of about 2% – depending on the data type – as compared to the same model without internal variables modelled with gs_param.

Validating Basic Checkpointing

To validate that we can indeed save and restore a SystemC subsystem inside Simics, we used a simple example device as shown in Fig. 3.6. This device exhibits all essential problems for checkpointing, namely state in memory-mapped registers and an `SC_METHOD` sensitive to an event that is posted at some point in the future using `sc_notify()`.

Or test device consist of a single memory-mapped register and a function sensible to a event. When the register is written, a periodic event is triggered at each fixed time (1 μs). This test device is integrated into a bare-bones PowerPC 603e-based machine in Simics, running a simple test program directly addressing device memory.

Our test consists of starting the simulation, doing a write to the register to start the periodic event, and taking a checkpoint. After taking the checkpoint, we quit Simics. Next, we start a new Simics session and resume the simulation

```
class systemc_greencheckpoint_test:
public sc_module,
...
SC_METHOD(function);
sensitive << my_event;
...
void
systemc_greencheckpoint_test::function()
{
   cout << "(SC code) function called at "
        << sc_time_stamp() << endl;
   my_event.notify(1, SC_US);
}
```

Figure 3.6. Code example

from the checkpoint. We could observe that the periodic event is triggering again as expected. With this test we validated that the SystemC kernel event list for SC_METHODs is properly saved and restored, and the simulation can be restored using our strategy.

Validating Model Updates

To validate the updatability of a model with new features using a checkpoint for an older version, we created another simple SystemC device model containing a single memory-mapped register. We then wrote a driver program for this device, executed the program to change the value of the register, and took a checkpoint. At this point, we exited the simulation, and changed the source code of the model to include an extra register. After recompiling, we started the simulation from the checkpoint without problems. The new register took on its default value as set in the SystemC source code, while the old register used the value provided in the checkpoint. We then executed the simulation for some more time, and took a new checkpoint. This checkpoint correctly contained the state of both registers, as affected by the software driver program (the driver knew about both registers from the start). Thus, we show that we can update models and use old checkpoints.

Checkpoint Size

Since the checkpoint system that we use only stores the essential data for a simulated system, it will generate compact checkpoints in general. For the validation example, the checkpoint was about 88 kB in size – most of which is the contents of the RAM of the simulated machine containing code. That can be compared to the overall process size of the simulation, at 263,688 kB, which also includes overhead such as the simulation core, simulation code, and user interface system. Using the "store memory contents to disk" approach to checkpointing would thus provide a checkpoint of some 256 MB, or 1,000 times larger. Saving the state of the system in a VmWare-style snapshot would have run to some 1.5 GB, for another factor of 6 in size.

Complex Test Case

Once the basic system was proven, a more complex test case was created. The SystemC device was a programmable periodic timer which could be reset, stopped, and started over a memory-mapped interface. It also had an outbound interrupt to the system interrupt controller. This device was inserted into a virtual Freescale MPC8641D-based system, which is a dual-core 32-bit Power Architecture machine running a Linux 2.6.23 kernel. The kernel was extended with a custom driver for our timer, and a user-level test program written to activate the timer over the driver. This test proved that our solutions works with realistic-scale systems, involving complex software, interrupts, and timed events in SystemC modules.

A screenshot of the system after a second Simics was started from the checkpoint is shown in Fig. 3.7. Note how the reported time and serial console state is the same in both simulations.

The checkpoint information for the SystemC subsystem in this example is shown in Fig. 3.8. Note that there is some Simics information about the attached interrupt device and the processor to post events on, in addition to the state of the SystemC system.

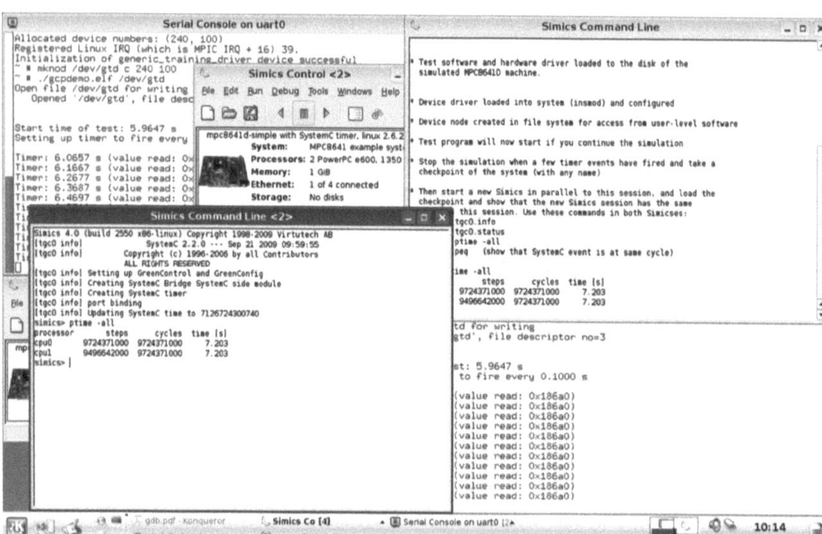

Figure 3.7. Complex checkpoint example

```
OBJECT tgc0 TYPE timer_greencheckpoint {
    queue: cpu0
    build_id: 0x9e2
    systemc_time: 0x5c10c637307
    max_time_skew: 0x174876e800
    irq_dev: (pic, "internal_interrupts")
    irq_level: 23
    irq_pending: 0
    gs_all_param_value: "timer_greencheckpoint.interrupt_status=0;
     timer_greencheckpoint.register_bell=1;
     timer_greencheckpoint.register_control=1;
     timer_greencheckpoint.register_countdown=100000;
     timer_greencheckpoint.register_status=0;"
    methods_events_list:
"6326694671111\n6426693333333;timer_greencheckpoint.timer_manager;"
}
```

Figure 3.8. Checkpoint information for the SystemC timer device

8. Discussion

This work has proved that it is possible to do checkpointing in SystemC, with a moderate effort and building upon existing frameworks. It has also exposed some issues in the SystemC design. The key problem is the provision of a Unix-style threaded execution model in SystemC, rather than an event-driven run-to-completion model. The threaded style encourages storing essential simulation state on the stack and as the location in the code, which is very hard to explicitly save and restore in a portable manner. Thank to the existence of SC_METHOD, a sound style can be implemented in the current SystemC framework. Note that no expressivity is lost by not using threads, as any threaded model can be easily translated to an event-driven model.

SystemC would have to be refined to make model state a first-class aspect of the language, and not just something implemented in arbitrary ways using C++ mechanisms as it is today. The connections between modules and the modules present in a simulation would also have to be first-class items. Finally, the kernel would need a host-independent representation of the state. Especially if state is to be exchanged between different SystemC kernels from different vendors, such standardization is needed.

An alternative way to enable checkpointing is to define a SystemC Virtual Machine and compile models to byte codes rather than to native code, thus providing a layer of indirection that can be used to dump and restore the system state without change to models. Such work has been done for Java, for example.

9. Conclusions

We have discussed the need and utility of checkpointing. It is an enabler for software development, and if it can be achieved in such a way as to allow the result to be made use of on different host platforms, it can be used in a number of ways.

In the past, checkpointing has been achieved by saving the complete process state of the process running the simulation. This limits the ability to distribute the simulation, and also is expensive in terms of disk and time.

Our aim has been to investigate the possibility of saving and restoring the SystemC kernel and model state itself.

In order to achieve this ambitious aim, we have had to limit ourselves in terms of what models we support. However, we have found, even with those limitations, that the results are helpful.

We are able to save and restore the state of SystemC methods, and the events which trigger them. We are also able to save and restore the states of models that are running on the SystemC simulator.

In the future we will be investigating other approaches to checkpointing, and dealing with the case of SystemC threads. For example, it could be possible to converting threads to methods using automatic tools [6].

References

[1] M. Bergqvist, J. Engblom, M. Patel, L. Lundegard, Some experience from the development of a simulator for a telecom cluster (CPPemu), in *Proceedings of the 10th IASTED Conference on Software Engineering and Applications*, November, 2006

[2] Cadence, SystemC save and restore Part 2 advanced usage. http://www.cadence.com/Community/blogs/sd/archive/2009/03/09/systemc-save-and-restore-part2-advanced-usage.aspx

[3] CoWare introduces first ever checkpoint/restart capability for native SystemC Virtual Platforms, April 14, 2008

[4] J. Engblom, *Simics system modeling*. Virtutech Whitepaper, May 2008

[5] J. Engblom, Modeling language produces TLM for virtual platforms. SCDSource.com (2008), `http://www.scdsource.com/article.php?id=166`

[6] R. Günzel, SCThreadConverter, `http://www.greensocs.com/en/projects/SCThreadConverter`

[7] N. Hardavellas, S. Somogyi, T.F. Wenisch, R.E. Wunderlich, S. Chen, J. Kim, B. Falsafi, J.C. Hoe, A.G. Nowatzyk, SIMFLEX: a fast, accurate, flexible full-system simulation framework for performance

evaluation of server architecture. ACM SIGMETRICS Perform. Eval. Rev. **31**(4), 31–35 (2004)

[8] P.H. Hargrove1, J.C. Duell1, Berkeley lab checkpoint/restart (BLCR) for linux clusters, in *Proceedings of SciDAC 2006*, June 2006

[9] W. Klingauf: Systematic transaction level communication modeling with SystemC, PhD Thesis, TU Braunschweig, Dept. of Integrated Circuit Design, 2008

[10] P.S. Magnusson, M. Christensson, J. Eskilson, D. Forsgren, G. Hallberg, J. Hogberg, F. Larsson, A. Moestedt, B. Werner, Simics: a full system simulation platform. IEEE Comput. **35**(2), 50–58 (2002)

[11] OSCI TLM2.0 User Manual, June 2008

[12] J.L. Peterson et al., Application of fullsystem simulation in exploratory system design and development. IBM J. Res. Dev. **50**(2/3), 321–332 (2006)

[13] R. Ramey, Boost serialization version 1.36 documentation, `http://www.boost.org`

[14] M. Rosenblum, M. Varadarajan, SimOS: a fast operating system simulation environment, Stanford University Technical Report CSL-TR-94-631, 1994

[15] C. Schröder, W. Klingauf, R.G.M. Burton, E. Roesler, Configuration and control of SystemC models using TLM middleware. CODES+ISSS '09

Chapter 4

EFFICIENT APPROXIMATELY-TIMED PERFORMANCE MODELING FOR ARCHITECTURAL EXPLORATION OF MPSOCS

Martin Streubühr, Jens Gladigau, Christian Haubelt, and Jürgen Teich
Hardware/Software Co-Design, Department of Computer Science, University of Erlangen-Nuremberg, 91058 Erlangen, Germany

Abstract In this chapter, we propose an efficient modeling approach that permits simulation-based performance evaluation of MPSoCs at Electronic System Level (ESL). The approach is based on a SystemC simulation framework and allows for evaluating timing effects from resource contention when mapping applications to MPSoC platforms. The abstraction level used for modeling timing corresponds to approximately-timed communication in transaction level models. This allows for an accurate performance modeling, including temporal effects from preemptive processor scheduling and bus arbitration. However, in contrast to standard OSCI TLM, application mapping and platform models are configurable and, thus, enable design space exploration at ESL. A Motion-JPEG decoder application is used to illustrate and assess the benefits of our approach.

Keywords: Performance modeling, Simulation, Electronic system level, SystemC

1. Introduction

The increasing computational demands of modern applications often require their implementation on heterogeneous multiprocessor platforms. Selecting the optimal platform and optimal mapping of the application to the selected platform is a challenging task. These decisions are made in early design phases, at the *Electronic System Level* (ESL). Design decisions made at this level have a huge impact on the performance of the final implementation, and therefore require a good knowledge of the design space to achieve the best possible result.

In practice, it is desirable to avoid costly redesign iterations until the performance requirements are met. From the practical point of view, this is especially

D. Borrione (ed.), *Advances in Design Methods from Modeling Languages for Embedded Systems and SoC's*, Lecture Notes in Electrical Engineering 63,
DOI 10.1007/978-90-481-9304-2_4,

true when redesign involves the lower abstraction levels, which require high effort. Modern design flows focus on moving the redesign iterations to higher abstraction levels [1], e.g., using Transaction Level Modeling [11]. As a consequence, an ideal top-down approach for system design looks like the decision pyramid in Fig. 4.1. Here, the design phase starts with a high level model of the application. To explore design alternatives, transaction level models are derived from the high level model. These TLMs are used for virtual modeling of the architecture. In early phases, a so called *loosely-timed* communication allows for fast simulation and helps in making coarse grained design decisions. For example, a selection of an appropriate number of microcontroller, IP-cores, and busses can be achieved. In later phases, promising design alternatives are further refined to so called *approximately-timed* models. Especially

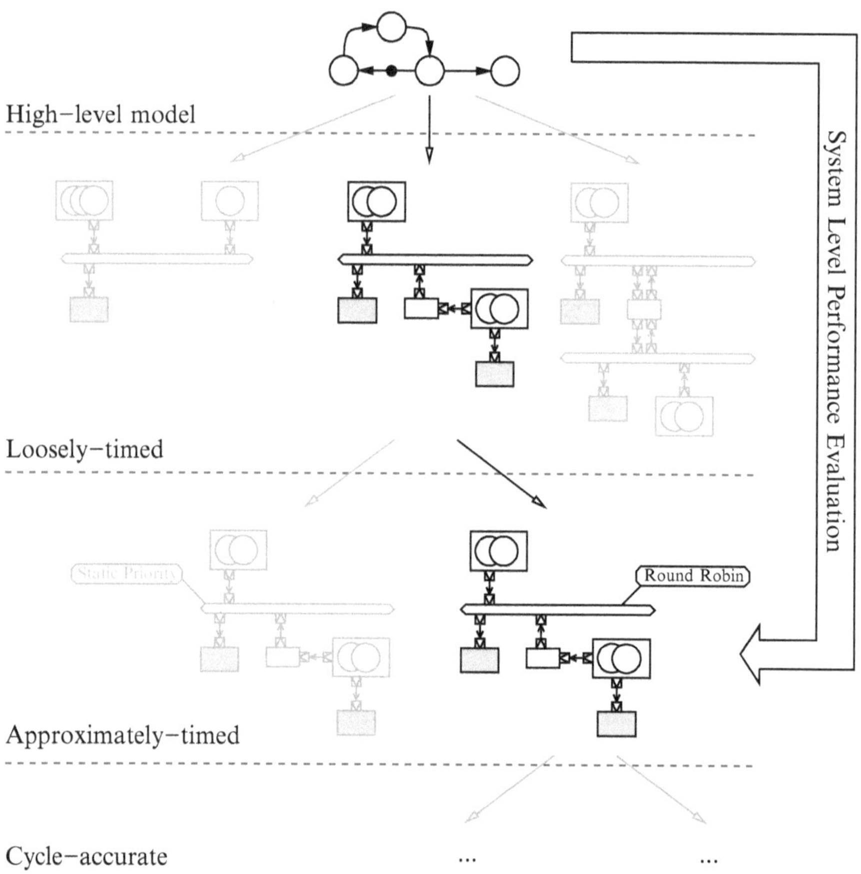

Figure 4.1. Decision pyramid: a high level model is refined stepwise to more detailed abstraction levels for performance estimation. Our approach applies system level performance estimation directly to the high level model and can be used as a short-cut to a concrete design

selecting communication architectures, arbitration, and scheduling policies, as well as task and communication priorities require more detailed performance simulations.

Such a hierarchical refinement process allows for reusing models from higher abstraction levels, in contrast to directly developing an RTL implementation from a specification. Nevertheless, at all abstraction levels several implementation alternatives are modeled for evaluation and comparison, and thus, the design-flow involves several iteration phases. Despite reusing large portions of those TLMs, each model requires manual modeling and implementation effort for evaluation. Due to limited resources and time-to-market pressure, the first design meeting the performance requirements looks invitingly beautiful for implementation. As a consequence, the final design may be over-constrained and far from ideal, as there may be some room left for optimizations not obvious to a designer.

We present an approach that supports the decision making phase using simulation-based performance evaluation. Our methodology is based on combined functional and performance simulation of a system level model. Thereby, design decisions, like architecture selection, task mapping, and transaction routing, as well as scheduling and arbitration policy selection, are configured in the performance model. As a consequence, the design space can be explored with significantly reduced effort, as, in contrast to compiled performance models, our performance model is configured at elaboration time. The abstraction level of communication for the performance modeling is comparable to the approximately-timed coding style of TLM. Our methodology can be used as a short-cut from the high level model to an approximately timed TLM, as indicated in Fig. 4.1. Thereby, it significantly reduces the cumbersome and error-prone manual implementation of design decisions. The selected design is used for further refinements towards the final implementation.

The remainder of this chapter is structured as follows: Related work is discussed in Section 2. Temporal abstraction levels defined in the OSCI TLM standard are discussed in Section 3. In Section 4, the input for automatic generation of the proposed System level Performance Model is discussed. In Section 5, the simulation-based performance estimation is described. Section 5 presents results from applying the proposed methodology to a multimedia application mapped onto MPSoC platforms. Conclusions are given in Section 7.

2. Related Work

The Sesame approach [2] is used for simulation of embedded architectures. A trace-driven co-simulation of application and architecture models allows for performance evaluation of different hardware/software mappings. The application, modeled as a Kahn process network, is co-simulated with an architecture model. Traces of application events from the KPN simulator, representing computation and communication, drive activations in the architecture simulator. The approach models no feedback from the architecture model to the application model. Thus, it achieves a good simulation performance, but is limited to deterministic application models, e.g., KPN. In contrast, for non-deterministic models, e.g., dynamic data-flow, the activation of tasks may depend on the (partial) execution order of other tasks. Thus, an activation-based performance modeling, i.e., feedback of execution times from the architecture model, is necessary for a correct combined functional and performance simulation.

A simulation approach that supports such a feedback is described in [4, 5]. It permits modeling of real-time scheduling strategies on the basis of SystemC. For this purpose, an extension of SystemC to model executable software tasks is proposed. The end of each atomic operation, e.g., each SystemC code line, is augmented by an `await()` function call. The possibility to perform preemption is given between atomic operations. In this approach, a set of software tasks is assigned to each processor. This technique provides the ability to simulate a set of processors running an operating system in parallel to hardware modules. Unfortunately, using this approach requires a substantial modification of the source code.

Another approach, described in [7], proposes a so-called *Virtual Processing Unit* (VPU) running several tasks using a priority-based scheduler. Software processes are modeled as *timed Communicating Extended Finite State Machines* (tCEFSM), where each transition represents an atomic operation. Moreover, each transition consumes the same fixed amount of processor cycles. In this approach, several tasks modeled as tCEFSM can be executed sharing the same VPU. Task execution times depend on the simulated VPU clock speed. The main limitation of this approach is their modeling of time, as each transition of a tCEFSM requires the same number of processor cycles, and the support of only a single priority-based scheduling strategy.

An approach for simulation of heterogeneous multi-processor platforms in SystemC is given in [12]. There, a task-accurate performance evaluation is applied to applications modeled in SystemC mapped to platforms containing multiple processors. Modeling of arbitrary scheduling policies is a special focus of this approach. In addition, our work is beyond modeling of task execution times, and allows for performance evaluation of multi-hop communication architectures using approximately-timed modeling.

In this chapter, we will overcome the limitations imposed on the discussed approaches by: (1) activation-based performance modeling, (2) transaction-based communication routing, (3) modeling of resource contention and preemption, (4) modeling of arbitration and scheduling policies, and (5) configurable task to resource mappings with configurable execution delays.

3. Temporal Abstraction Levels

The OSCI Transaction Level Modeling (TLM) standard allows for modeling bus-based communication in SystemC system level designs on different levels of temporal abstraction. Main aspects of the standard are:

- Separation of module functionality and module communication
- A common abstract interface, which allows for easy experimenting with different bus architectures, and reusing IP code that obeys the TLM standard
- Coding styles for different use cases

A transaction is an abstraction of communication, taking place between SystemC modules, e.g., a CPU reading from a memory. Transactions have an initiator and a target; between them, the transaction is transported using transport functions. The transaction may pass so called *interconnect components*, like busses or routers, on its way from initiator to target. Transport functions can be either blocking or non-blocking, i.e., are allowed to call or shall not call SystemC's wait function, respectively.

There are two coding styles defined in the TLM standard: (1) loosely-timed and (2) approximately-timed. Basically, these styles refer to different temporal abstraction levels for the communication. While in loosely-timed communication blocking transport functions are used, the communication in approximately-timed is implemented by non-blocking functions. See Fig. 4.2 for an illustration of both. There, three transactions are executed using the two coding styles.

Communication in loosely-timed models allows for exactly two timing points associated with each transaction, call and return of the blocking transport function, respectively. In loosely-timed communication, the transaction is processed within the context of the initiator of the transaction solely; context switches due to multi-hop communication are avoided. In Fig. 4.2a, this is indicated by a solid arrow for each transaction.

For more precise (i.e., more timing accurate) simulation of interleaving multi-hop communication, several timing points are needed per transaction. In Fig. 4.2b, this is indicated by discontinuous lines for the three transactions. This level is needed, e.g., to simulate congestion on busses, and to experiment with arbitration strategies. Communication on this abstraction level is called approximately-timed and is implemented using non-blocking transport functions.

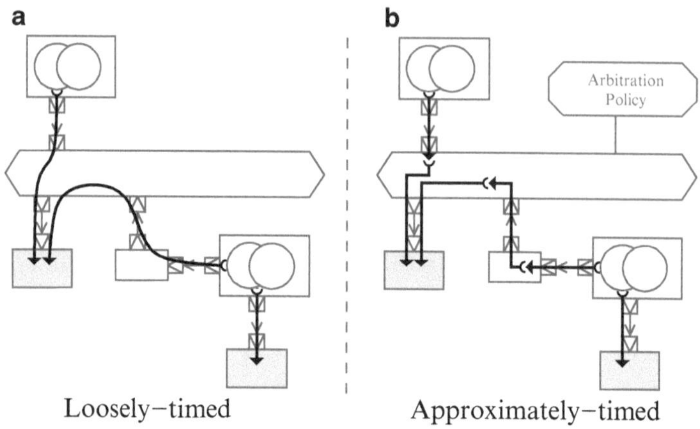

Figure 4.2. Loosely-timed transaction level models use blocking transport functions executed in the context of the initiator. Approximately-timed models use non-blocking transport functions. Context switching allows for a fine-grained timing simulation including effects of resource contention and arbitration

4. Modeling

Our performance modeling approach relies on the separation of functionality and architecture, represented by application and platform models. Thus, it is based on the Y-chart approach [8]. As illustrated in Fig. 4.3, the functionality is given by a set of applications written in SystemC. Tasks a_0 and a_1 represent SystemC modules with a single process. Communication between tasks is handled by SystemC channels. Additional platform applications may be modeled also, e.g., modeling the software provided by the platform vendor. The architecture represents the components and connections of the heterogeneous MPSoC platform and provides a technical basis for implementing the functionality. The selected platform is specified in a configuration file. Finally, a mapping constraint file assigns tasks from the applications to the components of the MPSoC platform. Given this information, a System level Performance Model (SPM) is automatically generated in SystemC as discussed below. As the SPM is implemented using SystemC, performance evaluation is done by analyzing the simulation trace. This is discussed in Section 5. By changing the platform configuration file and the mapping constraint file, different design options can be explored easily and fast. This architectural exploration can even be automated, cf. [6].

Application Modeling

To allow performance estimation, the user first specifies the application or a set of applications. For applying the proposed performance modeling approach, it

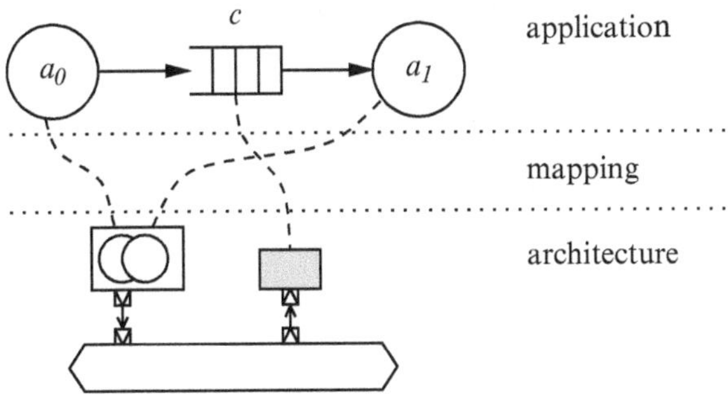

Figure 4.3. A data-flow model is mapped to an MPSoC platform. Actors are mapped to computational resources like CPUs. Communication channels are mapped to memories

is a prerequisite that an application is composed of a set of coarse-grained tasks, allowing task level parallelism, i.e., several SystemC modules with a single thread. Furthermore, each task shall be composed of atomic execution phases. Communication between tasks may occur only at the beginning or at the end of execution phases. Moreover, communication shall take place on dedicated communication channels only, and no implicit communication, like global memory, shall be used. This explicit modeling of communication is required for mapping tasks to resources without side-effects.

For the proposed methodology, we use actor-oriented modeling in SystemC. The concepts of actor-oriented design for embedded systems [9, 10] guarantees that the functionality is defined in a way such that it is independent from the actual mapping to an MPSoC platform [6]. Actors model concurrent parts of the application and are restricted to communicate through dedicated channels only. Thus, a set of actors can be partitioned into hardware and software.

We use a SystemC library [3] for actor-oriented design. The library provides special channels for inter-actor communication, and a C++ syntax to specify the actor behavior by means of a finite state machine. Channels are queues with FIFO semantics, allowing consumption/production of data in a first-in, first-out discipline. In the context of SystemC and TLM, actors correspond to tasks, and read/write accesses to channels correspond to transactions.

Modeling MPSoC Architectures

The MPSoC platform is modeled by a set of components, representing microcontrollers, DSPs, hardware accelerators, memories, and busses. Each component is characterized in the platform configuration file by a scheduler, required to resolve resource contention, and a *transfer delay*. This transfer delay is the time needed to transport a single byte of data from one component to another.

The platform characterization is done using an XML file. A part of this XML file is shown below:

```
<platform>
  <component name=''CPU''>
    <scheduler type=''Round Robin''/>
    <transfer delay=''0.83 ns''/>
  </component>
  <component name=''Bus''>
    <scheduler type=''Priority''/>
    <transfer delay=''2.74725 ns''/>
  </component>
  <component name=''Memory''>
    ...
</platform>
```

Within the platform tag, the components of the MPSoC platform are defined. In the above listing, e.g., a component CPU is specified with a round robin scheduling policy. Other components of the platform, including hardware accelerators, busses, and memories, are characterized in a similar way.

Setting the Mapping Constraints

Mapping constraints are given also via an XML configuration file. The configuration provides two sorts of information, namely the mapping of actors and channels to components, as well as the routing for transactions. Beside these, estimated computation times for actor execution on the specified component are annotated to the mapping information. The transaction routing defines the routes for each actor's read and write access to or from a channel. This implies that the components are required to handle a memory access.

A fraction of the XML file is shown below:

```
<mappings>
  <mapping actor=''a_0'' component=''CPU''>
    <delay action=''sendData'' value=''2,032 ns''/>
    <delay action=''newScan'' value=''1,518 ns''/>
  </mapping>
    ...
</mappings>
<topology>
  <route src=''a_0'' snk=''c''>
    <hop name=''CPU''/>
    <hop name=''Bus''/>
    <hop name=''Memory''/>
  </route>
  ...
</topology>
```

Actor a_0 is bound onto component CPU. Executing action *sendData* requires 2,032 ns in this configuration. Moreover, the data written to channel c always passes the components CPU, Bus, and Memory.

5. Performance Evaluation

Beside the separation of functionality and architecture, our performance evaluation approach is characterized by two basic concepts: (1) We use an activation-based modeling of task execution times including temporal effects of preemption and task scheduling that may occur in embedded processors. (2) We model transaction-based communication. Communication delays are modeled for transactions, incorporating effects of multi-hop communication and bus arbitration policies, i.e., the abstraction level corresponds to approximately-timed TLM.

Given the set of actor-oriented applications written in SystemC, the platform characterization, and the mapping constraints, a System level Performance Model (SPM) is generated automatically. For this purpose, the application is linked with a library providing so called *Virtual Processing Components* (VPCs). A VPC is a SystemC module with an event-based scheduling interface. For each component in the platform characterization, a VPC is generated at elaboration time, and a scheduler is instantiated according to the configuration file. Different MPSoC platforms and application mappings can be evaluated by simply changing the configuration files.

Activation-Based Execution

For a combined functional and performance simulation, we need to distinguish the phases of activity of an actor, namely communication phases and execution phases.

The SystemC library for actor-oriented modeling [3] uses a finite state-machine to model the control flow of an actor. Identifying the execution phases is straight forward. Each transition in the actor's FSM corresponds to a phase of reading input data, a phase of executing the actor's functionality, and a phase of writing output data. Each actor execution phase is disclosed to the VPC by a simple function call, blocking the actor for further execution during performance simulation. As a consequence, each execution phase in actors is simulated in lockstep with the corresponding performance simulation. Timing simulation includes possible overhead from preemption, scheduling, and arbitration. For example, in Fig. 4.4, a simulation trace from applying a

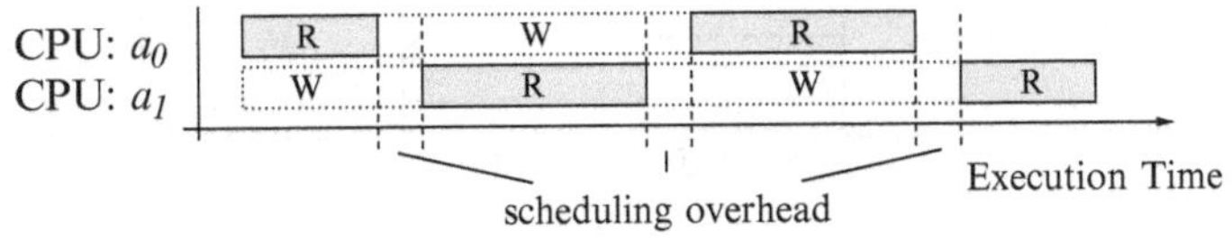

Figure 4.4. Two actors a_0 and a_1 are mapped to a CPU. A scheduling policy applies a fair round robin schedule. Thus, execution of actors alternates

preemptive round robin scheduling policy is depicted. It can be seen that actors a_0 and a_1 are executed alternating. Whenever the time slice of the round robin scheduling is expired, another actor is selected for execution. Additionally, scheduling introduces some overhead, which is respected in simulation.

Transaction-Based Communication

Communication between actors in the application model is limited to dedicated channels. Those channels serve the purposes of data exchange and synchronization. Implementing such a channel in MPSoC platforms requires mapping logical data storage to physical memory. Furthermore, transactions for writing and reading data and transactions for synchronization between actors need to be routed through the MPSoC. A generic transaction-based communication pattern for abstract communication channels is given in Fig. 4.5. This pattern consists of a memory, as well as a read, a write, and two synchronization transactions. Every read/write in the data-flow model results in one or more data and one synchronization transactions.

For example, a queue with FIFO semantics, as used in a data-flow model, stores a fixed number of *data tokens* in order of insertion. The availability of tokens at the reader's end of the queue and free space at the writer's end of the queues may induce the activation of actors. Mapping the data-flow model to an MPSoC platform requires mapping the data storage of the queue to memories in the platform model. Push and pop operations to the queue require to write to memory or to read data from memory and to send synchronization transactions.

In the same manner as in execution phases, the communication transactions are disclosed to the VPC library. The performance simulation is started, whenever a communication transaction is executed in the application. Timing

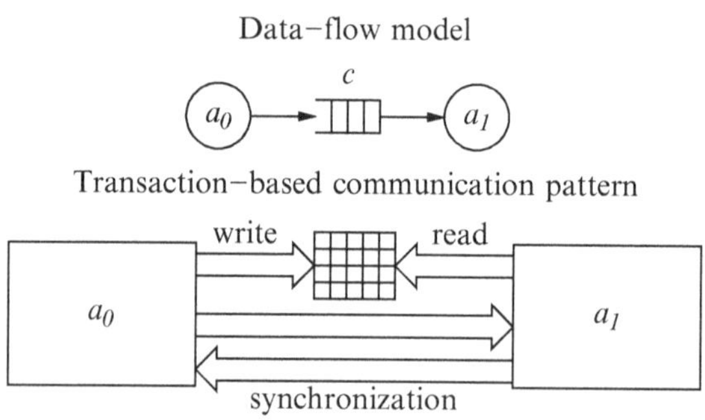

Figure 4.5. A generic transaction-based communication pattern is derived from a data-flow high level model. Push and pop operations to a FIFO queue are represented by read, write, and synchronization transactions

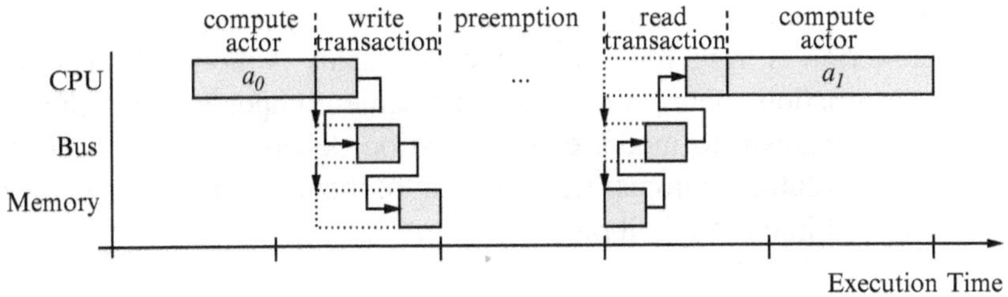

Figure 4.6. The hops of a multi-hop route are blocked for transactions representing memory access operations

simulation for the transaction routing potentially involves several hops along a route. Additionally, overhead from arbitration and preemption may occur at each hop.

Especially to model memory access, we need blocking communication, as depicted in Fig. 4.6. E.g., the write transaction from actor a_1 in Fig. 4.6 corresponds to a memory access. All hops along the route are blocked at the beginning of the transaction. In the example, the CPU is stalled and afterwards successively the Bus and the Memory are locked. When the entire route is locked, the delays for transmitting the data along the route are simulated. Locking of resources may only start at hops allowed to initiate memory transactions, e.g., CPUs and IP-cores. After locking the entire route, timing simulation is started at the initiator hop (CPU) for a write transaction, and at the target hop (Memory) for a read transaction, corresponding to the forward path and backward path of the transactions.

Performance Evaluation

During elaboration of the SystemC application model, the VPC library is configured with the mapping constraints and, hence, the estimated computation times for each actor according to the given mapping. This information is used during simulation to assess the performance (execution time) of the system. Each time an actor's execution phase is performed, its associated execution time is simulated. Before consuming or producing any data, the actor notifies the component the actor is bound to. Actor execution is blocked by this notification. Then, the component add's the actor to a ready list of the associated scheduler. The scheduler decides when an actor should be scheduled and, thus, computes when the computation time of the execution phase expires. This concept is similar to [12]. As the interface between VPC and the scheduler is event-based, preemptive and non-preemptive scheduling policies can be implemented.

Simulation of communication delay follows the very same principles. Every time a transaction is triggered, communication delay is modeled for all hops along the transaction route. We can use the same components and the same scheduling mechanism to model communication delays and arbitration policies. Multi-hop routing is implemented on top of the performance modeling of components within our VPC library.

6. Case Study: Motion-JPEG Decoder

In this case study, the proposed performance modeling is applied to a multimedia system. In particular, a Motion-JPEG decoder application is mapped onto a multimedia architecture template. The actor execution times are derived from another case study investigating the codesign of a Motion-JPEG decoder on Xilinx FPGA platforms [6].

Design Space Exploration

The application model of the decoder is specified in actor-oriented SystemC code using 21 actors and 56 channels. An architecture template incorporates 12 programmable processing elements and communication resources, as well as 10 dedicated hardware accelerators, connected via 36 point to point communication resources including storage memory. More concrete, the MPSoC platform may include two ARM9 and two ARM11 processors-cores, one DSP, three memories, two busses, and one gateway connecting these busses.

A sequence of images in SQCIF-format (128×96 pixels) is used as input stimuli for performance simulation. With the help of an automatic design space exploration, we generated 7,600 different mappings for the Motion-JPEG decoder. The performance evaluation of all 7,600 mapping variants took about 17 h in total. Note that three implementations provide a throughput of more than 24 fps and, thus, fulfill current real-time video requirements.

Transaction Delays

In a second experiment, we took a detailed look at the delay of individual transactions. The motivation is to assess the difference between approximately-timed simulation and loosely-timed simulation. Note that differences in communication delays between approximately-timed and loosely-timed modeling can only occur if congestion occurs on busses.

We use a mapping from the first experiment and the very same sequence of images as test case. For each kind of transaction, we recorded the individual communication latencies. This simulation setup has 44 active communication channels and we recorded the *read* and *write* transactions to the memories only. Figure 4.7 depicts the minimum, maximum, and average communication delay for those 88 transactions. About the half of the number of transactions has

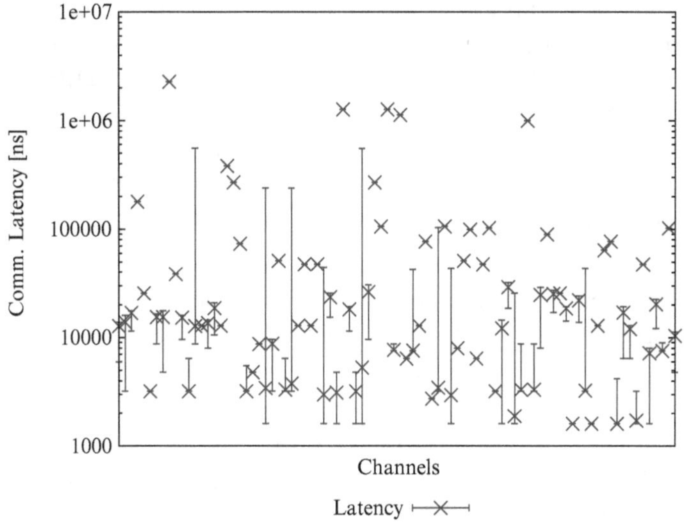

Figure 4.7. Minimum, maximum, and mean values for the communication delay for read and write transactions are recorded during performance simulation. Due to approximately-timed modeling, effects from resource contention and arbitration cause variable delays

fixed communication delays in this simulation. For the remaining transactions, we observed significant variations in the communication delay due to effects from bus arbitration and memory stalls. Thus, the approximately-timed modeling of those transactions, using our approach, is necessary for at least those transactions, for sufficient performance estimation.

7. Conclusions

Design decisions made at Electronic System Level require a good understanding of the design space. We presented a simulation-based performance evaluation framework for MPSoC platforms. Approximately-timed modeling allows for accurate performance estimation, even at a high level of abstraction. In our approach, platform model and mappings are configured at elaboration time and, thus, allow for fast design space exploration at ESL.

References

[1] R. Atitallah, S. Niar, S. Meftali, J.L. Dekeyser, An MPSoC performance estimation framework using transaction level modeling, in *Proceedings of the International Conference on Embedded and Real-Time Computing Systems and Applications* (IEEE Computer Society, Daegu, Korea, 2007), pp. 525–533

[2] C. Erbas, A.D. Pimentel, M. Thompson, S. Polstra, A framework for system-level modeling and simulation of embedded systems architectures. EURASIP J. Embed. Syst. 2007:Article ID 82123, 11 (2007). doi:10.1155/2007/82123

[3] J. Falk, C. Haubelt, J. Teich, Efficient representation and simulation of model-based designs in SystemC, in *Proceedings of the Forum on specification & Design Languages*, Darmstadt, Germany, 2006

[4] P. Hastono, S. Klaus, S.A. Huss, An integrated SystemC framework for real-time scheduling assessments on system level, in *Proceedings of International Real-Time Systems Symposium (RTSS)*, Lissabon, Portugal, 2004

[5] P. Hastono, S. Klaus, S.A. Huss, Real-time operating system services for realistic SystemC simulation models of embedded systems, in *Proceedings of Forum on specification & Design Languages*, Lille, Frankreich, 2004, pp. 380–391

[6] J. Keinert, M. Streubühr, T. Schlichter, J. Falk, J. Gladigau, C. Haubelt, J. Teich, M. Meredith, SYSTEMCODESIGNER – an automatic ESL synthesis approach by design space exploration and behavioral synthesis for streaming applications. Trans. Design Automat. Electron. Syst. **14**(1), 1–23 (2009)

[7] T. Kempf, M. Dörper, R. Leupers, G. Ascheid, H. Meyr, T. Kogel, B. Vanthournout, A modular simulation framework for spatial and temporal task mapping onto multi-processor SoC platforms, in *Design Automation & Test in Europe*, Munich, Germany, 2005, pp. 876–881

[8] B. Kienhuis, E. Deprettere, K. Vissers, P. van der Wolf, An approach for quantitative analysis of application-specific dataflow architectures, in *Proceedings of the International Conference on Application-Specific Systems, Architectures and Processors*, Zurich, Switzerland, 1997, pp. 338–349

[9] E.A. Lee, S. Neuendorffer, Actor-oriented models for codesign: balancing re-use and performance, in *Formal Methods and Models for System Design* (Kluwer, Norwell, MA, 2004), pp. 33–56

[10] E.A. Lee, S. Neuendorffer, M.J. Wirthlin, Actor-oriented design of embedded hardware and software systems. J. Circ. Syst. Comput. **12**(3), 231–260 (2003)

[11] Open SystemC Initiative (OSCI) TLM Transaction Level Modeling Library, Release 2.0 (2009) http://www.systemc.org.

[12] M. Streubühr, J. Falk, C. Haubelt, J. Teich, R. Dorsch, T. Schlipf, Task-accurate performance modeling in SystemC for real-time multi-processor architectures, in *Proceedings of Design, Automation and Test in Europe* (IEEE Computer Society, Munich, Germany, 2006), pp. 480–481

Chapter 5

FAST SYSTEMC PERFORMANCE MODELS FOR THE EXPLORATION OF EMBEDDED MEMORIES

Hans-Peter Loeb[1] and Christian Sauer[2]

[1] *Heinz Nixdorf Institute, Paderborn, Germany*
loeb@ieee.org

[2] *Cadence Design Systems, Munich, Germany*
sauerc@cadence.com

Abstract The quantitative exploration of the memory design space is needed early in the design process of deeply embedded systems. For predictive results, models are required that carefully consider three performance-impacting effects of memory accesses: synchronization, arbitration, and latency. Our approach combines functionally-correct performance simulation with memory models of different accuracy. Extending a SystemC-based framework, we leverage performance annotations to capture dynamic effects in the system. Indicative performance figures can be derived using only sparse synchronization and a statistical memory access distribution model. The models are shown to accurately reflect design changes while running up to four orders of magnitude faster than RTL simulation. A case study demonstrates our approach exploring tradeoffs between local and shared memories in a multiprocessor SoCs using a Wireless LAN packet processing application.

Keywords: SystemC, Performance modeling, Memory architecture, Embedded systems, Design space exploration, Wireless LAN

1. Introduction

The memory hierarchy of an application-specific embedded system significantly impacts its resulting performance. Thus, its key properties and the interplay with the application must be explored and defined early in the design process to guide the implementation of efficient SoCs.

D. Borrione (ed.), *Advances in Design Methods from Modeling Languages for Embedded Systems and SoC's*, Lecture Notes in Electrical Engineering 63,
DOI 10.1007/978-90-481-9304-2_5, © Springer Science+Business Media B.V. 2010

A typical early performance/area tradeoff in embedded systems (cf. Section 2) is between local and shared on-chip memories. On the one hand, local memories can be accessed quickly and are always available but come at the price of limited flexibility and extra area. Shared memories, on the other hand, are flexible in their usage and typically require (for the same total size) less area. However, they introduce access penalties that are difficult to estimate due to the effects of latency, arbitration, and multiple clock synchronization. Thus, memory models are needed that reflect these penalties as precisely as possible. These models must be fast since vast numbers of design points must be evaluated due to multiple design axes and large parameter ranges.

For a predictive estimate of the resulting system, its full function must be considered during the exploration. Simulation based on functionally-correct and timed performance models is required to fully expose the complex dynamic interactions, e.g. in real-time protocol processing.

Performance modeling frameworks aim at the fast exploration of design points. They annotate the target platform's performance using tags and execute the system function natively on a simulation host (e.g., in [1]). This effectively decouples execution from performance estimation but makes explicit modeling of the memory hierarchy necessary.

Our work extends the SystemClick performance modeling framework [2] for the exploration of memory hierarchies. The modified framework incorporates memory models at different abstraction levels, synchronization granularity, and access schemes. In particular, the contributions of this chapter are:

- A memory exploration framework that flexibly considers memory latency, arbitration, and synchronization effects.
- A memory model predicting the impact of contention occurring for groups of accesses, either in worst or average case, which improves prior-art significantly.
- A realistic case study from the wireless domain, which underpins the usefulness of our approach for evaluating large numbers of design points quickly.

The framework is tailored to the embedded memory design space as outlined next. Prior art on modeling and exploration of memories is discussed in Section 3. Section 4 then describes the SystemC framework focusing on the new memory modeling facet. Section 5 continues with detailed models for memory accesses. Their quality is examined in Section 6, and an exploration case study follows in Section 7.

2. The Memory Design Space

We are interested in exploring memory hierarchies for programmable SoCs as they are used for network and protocol processing in devices such as home gateways and network processors. Such deeply embedded systems have specific characteristics to fulfill their performance and real-time requirements:

Processing elements (PEs) deploy small RISC-like processing cores. Their operating frequency varies with the application. The PE subsystems comprise instruction and data memories running at core speed. Caches are avoided due to their unpredictable timing. Local memory content is loaded explicitly.

The numbers of deployed PEs range from one to the 10s. Each of them may be configured individually in terms of type and frequency, local storage, and communication and memory ports. Inter-PE communication (e.g., by message passing) occurs directly or via shared memory.

Shared memories are primarily SRAM and kept on-chip for cost and performance reasons. They may be organized in multiple banks and are accessed via bus/crossbar structures. Their operating frequencies differ from the PE frequencies. Thus, accesses require synchronization in addition to arbitration.

Other clients, such as on-chip CPUs or DMA engines, may access selected shared memories as well, effectively increasing the contention and limiting the bandwidth.

The setting in this paper reflects these characteristics: For a Wireless LAN application, the system has two PEs, a crypto coprocessor, and three DMAs to handle IO and the accelerator. The message passing communication is direct, and we consider local and shared memories at various speeds and distances.

3. Related Work

Analytical models based on queuing networks or Petri nets have been used to reflect the impact of memory arbitration and latencies in large-scale systems. They typically have difficulties reflecting the control flow and different request rates. This has lead to very complex [3] or overly simple solutions [4]. The latter does not cover contention between only two clients and asymmetric request loads. In contrast, there is an industry trend toward accurate co-simulation of MP-SoCs, e.g., coupled at TLM abstractions. While processor models relying on binary translation or performance annotation frameworks [1, 5] have reached a sophisticated level, they require fine-grained simulation synchronization and offer per se no exploration methodology.

Closer to our work are simulation approaches abstracting shared resource accesses. Bobrek et al. [6] combine a proprietary discrete event scheduler with statistical models. We extend on this by presenting an improved scheduler built upon SystemC. Also, their approach uses the formula from [4] and thus lacks accuracy for our problem at hand. This is improved in [7], where contention estimations are based on access attributes derived from initial co-simulation using regression learning. For this, a cycle-accurate system is required, which is difficult early in the design process and when the memory system itself is subject to exploration. The authors report dependencies on the initial simulation and give results for homogeneous benchmark kernels.

At the system level, approaches such as [8] focus on real time behavior in prioritized systems to iteratively derive worst-case execution times from abstract stimuli. Similarly, [9] combine formal network calculus with simulation. In [10], a trace is used to derive a communication graph from co-simulation. While complex architectures are reflected, the control flow is limited to the simulation trace and may not depend on dynamic effects. Also a static task graph is used in [11] to synthesize memory hierarchies using an ILP approach.

In [12], memory optimization techniques are presented. There is a large body of work on loop optimization and memory design limited to algorithmic kernels, see, e.g., [13], and its impact on cached systems [14].

None of these approaches is suited to cover the design space of deeply embedded systems, to reflect complex application behavior, and to provide an easy-to-use methodology for efficient early design exploration at the same time.

4. Exploration Framework

We extend SystemClick [2] by an explicit memory modeling and exploration facet. SystemClick uses independent descriptions of the system function and the platform architecture, enabling seamless exploration of different mappings from functions to hardware blocks. The tool generates SystemC performance models (including function) from high-level descriptions employing library elements for applications and hardware resources (cf. Fig. 5.1). An analysis step provides feedback for optimizing architecture, application, and mapping.

System Function in Click The application, i.e., the functionality, is specified in an architecture-independent way using *Click*. These models capture the concurrency in the application and dependencies naturally. Communication is explicit and described by the flow of packets in the model. The system function is verified using a simulation engine and traffic generators.

Architecture Models in SystemC Architecture building blocks are specified as performance models that represent shared computation, communication, and storage resources. Each building block is described by a resource manager (RM) in SystemC that arbitrates tasks and consumes time. RMs also specify the properties of a resource, such as its type, operating frequency, and scheduling policy. Abstract RM elements can be inferred automatically to speed up the exploration process. SystemC enables step-wise refinement, mixed-level simulations, and virtual prototyping.

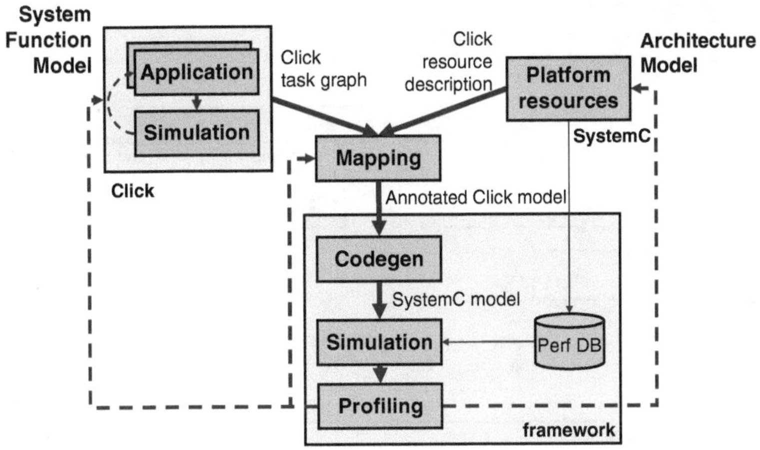

Figure 5.1. Platform exploration with SystemClick

Mapping and Code Generation The application graph is partitioned at element boundaries and mapped onto an SoC platform manually. To ease this task, specialized Click elements are provided that guide the code generation process. Each task of the partitioned graph is associated with a computation resource; its data objects with storage resources; connections between tasks with communication resources. Additional wrappers represent the operating system aspects of a platform. From the annotated source, the framework generates a SystemC model that is compiled together with libraries for application and architecture using standard tools. Using a performance database for elements and subfunctions, behavior and timing of the full system can be simulated.

Representation of Data and Storage

Data objects in the functional model are described using regular C/C++ constructs. They are tagged with a memory ID to allow access profiling and mapping onto a platform instance. Depending on the desired mapping and analysis resolution, a memory ID may identify individual objects or groups of objects to be allocated into the same memory area. In the mapping phase, each memory area is associated with a storage resource of the platform. Memory areas for common heap, stack, and compile-time data sections are created by default. The user may add extra heaps or compile-time sections.

Performance Simulation

The performance simulation executes tasks within a concurrent SystemC environment provided by wrappers (cf. Fig. 5.2). These handle a task's communication, interface it to the shared computation and storage resources, and

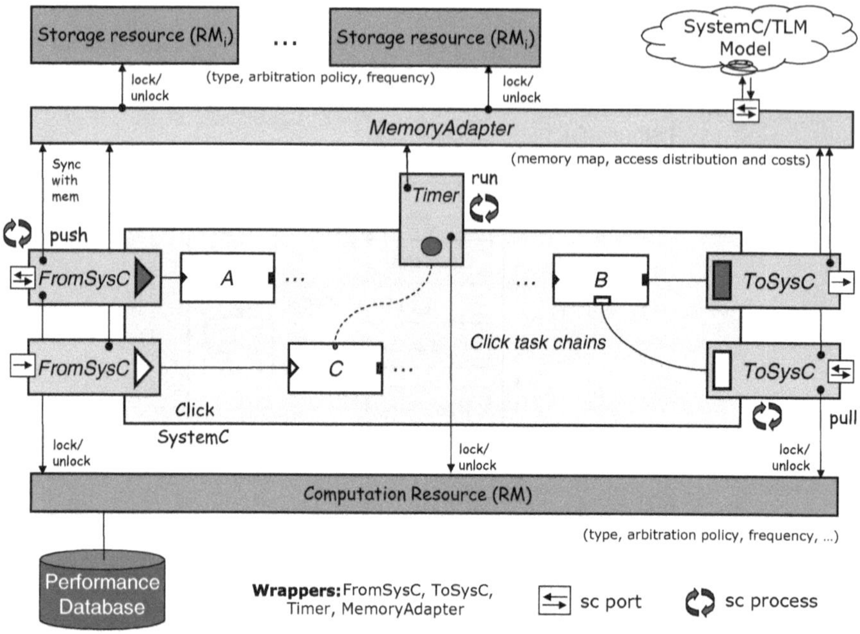

Figure 5.2. SystemC performance model for tasks mapped to one PE, represented by the computation resource, wrappers, and an adapter to associated storages

synchronize its processing time with simulation time. A performance database associates each task with costs for a particular resource, e.g., its processing time and the number and type of accesses to different memory areas. Using performance tags, costs are annotated and accumulated at runtime. A typical Click element has three to five of such performance tags. Performance data can be derived, e.g., by tag-wise profiling on embedded cores for predictive and cycle-precise models.

Depending on the simulation mode, synchronization between processing and simulation time may only take place at points of IO to minimize simulation events. Accumulated memory accesses within such a synchronization interval may be handled individually (single access mode) or as a group (grouped access mode) per storage resource to derive memory contention and access penalties (cf. Section 5). For increased simulation resolution, performance tags can seamlessly be used as additional synchronization points, e.g., in Fig. 5.6.

Analysis of a Memory Hierarchy

Building on the statistics and trace capabilities of resource managers, which trace idle/service utilization per task, specific performance data is gathered on (1) the accesses and stalls per processing resource and memory area, and (2) on the back log and waiting times at arbiters. Combined with other results such as throughput, response delay, and dynamic memory usage this data is the input for an analysis as shown in Section 7.

5. Memory Models

Performance models for memory accesses must be indicative of the total delay incurred by a requesting client. Figure 5.3 summarizes dynamic and static effects that need to be considered:

- **Synchronization** The local clock needs to synchronize with the memory subsystem, for example, through an asynchronous FIFO.
- **Arbitration and Contention** Identifying the next client to be granted access to a memory typically takes one clock cycle of the access system and can be based, e.g., on round-robbin.
- **Memory Latency and Delays** Static delays for request, memory access, and response are incurred. For read requests, the round trip delay becomes important and the subsystem, e.g., may remain unavailable until completion.

Integrating such models in SystemClick, the accumulated computation time μ_{comp} and memory accesses at a synchronization point are known prior to the actual synchronization with simulation time. After associating memory area accesses to storage mem_i using a memory map associated with every computational resource, the tuple

$$\mu = (\mu_{\mathrm{comp}}, \{\mu_{\mathrm{mem,i}}\}), \quad \mu_{\mathrm{mem,i}} = (\mathrm{acc}_{\mathrm{i,read}}, \ \mathrm{acc}_{\mathrm{i,write}})$$

is presented to the memory adapter for synchronization.

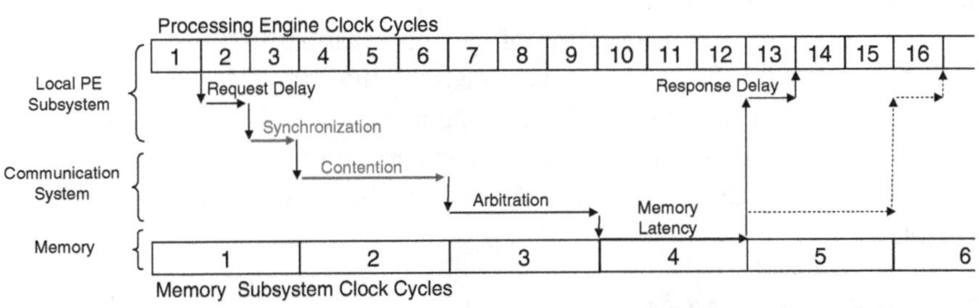

Figure 5.3. Examplary read request to a shared memory. Dynamic effects caused by synchronization and contention occur in addition to static latencies and delays

This allows two different approaches to reflect the latency of shared memory accesses: Accesses can be issued one-by-one to a representation of the architecture. Or, all accesses for the current interval can be grouped, allowing to predict additional contention penalties incurred, e.g., on a statistical basis.

Single Access Model

In single access mode, the load/store requests for all memories acc_{total} are ordered by the memory adapter (MA), e.g. using a Deficit-Round-Robin schedule to guarantee an even distribution. Every single access is issued either to an abstract resource manager RM_i representing storage mem_i or passed to the SystemC environment, e.g., to a TLM model (cf. Fig. 5.2).

Each access is divided into five steps, where steps 1–3 can be replaced by a single call to a SystemC port:

1 Call the blocking *lock()* function of RM_i.

2 *wait()* for the arbitration period of RM_i.

3 Call *unlock()* function of RM_i.

4 Account for extra delay associated with mem_i.

5 Account for a fraction of the total computation time.

Arbitration period and extra delay can be resource and load/store specific and may consist of local PE and memory cycles (see Fig. 5.3).

We implement a number of random, fixed, and trace-based distributions in the MA to determine the proportion of μ_{comp} in between two subsequent accesses. For example, a common approach in literature is to assume a negative exponential distribution, where we choose the parameter $\lambda = \mu_{\text{comp}}/acc_{total}$. In order to guarantee the total bound given by μ_{comp}, the rounding of cycles is balanced according to an updated λ, and accesses are squeezed if the end of the interval is approaching.

For cycle-accurate consideration of synchronization effects between differing local and memory clocks, it is necessary that the RM_i storages have a notion of time. For this, the RM_i defer the arbitration of the next client to the SystemC *update()* phase at the end of their current clock cycle. This synchronization is demand driven and only necessary when asynchronous requests are present.

Grouped Access Model

In this mode, all accesses to given storages mem_i during an interval $I = \mu_{\text{comp}}$ are grouped together and evaluated by special statistical RM_i's. Latencies for

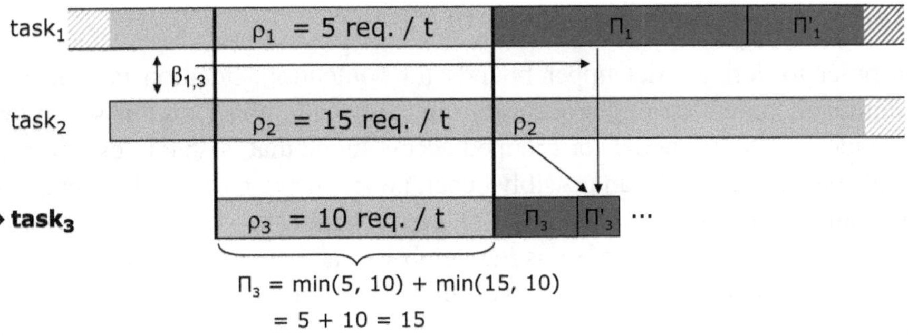

Figure 5.4. Example of estimating contention during an interval I in grouped mode. Penalty phases Π_j account for additional latency due to delayed accesses

memory accesses are abstracted in the computation time μ_{comp} as accumulated by performance tags. The effect of clock synchronization between PEs and memories is also considered statically, e.g., in the average or worst case. However, delay caused by memory contention is reflected in one (as in [6]) or more penalty intervals Π following the initial interval I, as shown in Fig. 5.4. With prerequisites from above, we execute four phases at a synchronization point:

1 Set client request rate $\rho_j = \mu_{\text{mem,i}}/I$ and interval I at all RM_i.

2 *wait()* the interval time I.

3 Enter penalty phase: Repeatedly retrieve and *wait()* penalties Π_i until remaining penalty is below threshold.

4 Clear client request rate at RM_i if no penalty left.

Statistical RM_i keep track of their utilization ρ, the set of active clients C_{act}, and requested utilizations ρ_j of every active client $c_j \in C_{act}$. Only when request rates are set, cleared, or penalties retrieved, the state is updated by calling an *updatePenalties()* function that computes penalties for the passed interval J a posteriori, e.g., using a statistical contention function. Of course, this assumes memory accesses and requests served to be equally distributed in any given subinterval. Setting the contention function to zero, only memory latencies are simulated, yielding lower bound execution times.

During penalty phases, clients iteratively consume, i.e., delete and *wait()*, their current penalty Π without clearing the requested utilization: This allows to consider contention for other clients caused by deferred accesses. Without iterating the penalty phase, the approach presented in [6] is covered, which ignores contention during penalty intervals.

Iterated Contention Algorithm

In order to derive strict upper bounds for contention, iteration is needed as additional contention might occur during the penalty phase. In this section we provide a memory model for grouped access mode that is able to estimate the total contention which can possibly occur, i.e., in the worst-case. The scheduler is arbitrary but assumed to be fair, i.e., every client is served at least $1/n$ of its request utilization ρ_j, where n is the number of active clients $|C_{act}|$ for I. The case of prioritized schedulers is well understood and out of the focus of this work.

In a first approach (S), we exploit scheduler fairness to derive the number of totally outstanding requests α_j of client c_j in the worst case: Initially, α_j is $\rho_j \times I$. For any interval J, α_j can be reduced by at least $\rho_j \times J \times (1/n)$. Now, the penalty Π_j equals the number of outstanding requests α_j and has to be recomputed for every subsequent interval. The approach, however, leads to overly pessimistic bounds and performs badly due to many iterations of update intervals.

For an improved approach (P), we first observe that the worst-case penalty incurred by any client $c_j \in C_{\text{act}}$ in any interval J is given by

$$\Pi_j = \left(\sum_{c_k \in C_{act}, c_k \neq c_j} \min(\rho_j, \rho_k) \right) \times J. \tag{5.1}$$

Any request made by c_j in J can only collide at most once with every other client, again, using scheduler fairness (see also Fig. 5.4).

It is important to note that all possible contention between two clients c_j and c_k is already considered prior to their penalty phases. Thus, the penalty at c_j incurred during c_k's penalty phase cannot be larger than the penalty c_k already had when c_j became active. In this case, we can bound the partial penalty $\pi_{j,k}$ by the a priori penalty $\beta_{j,k}$, which is stored for all other clients c_k whenever a client c_j becomes active.

Both methods are independent and can be combined, as sketched in Fig. 5.5: The penalty is computed using Eq. 5.1 and bound by both a priori penalty and remaining accesses. The worst-case penalty derived by the presented algorithm is a good indication of the actual amount of contention that may occur. For an estimate of the average case, this penalty can be scaled in each interval J, e.g., based on $|C_{act}|$, ρ, ρ_j, or the burstiness of the client. The *grouped avg.* approach simply divides the penalty by the number of active clients in the given interval, effectively distributing contention evenly amongst the clients.

```
for all active clients c_j ∈ C_act
  update outstanding requests α_j                    /* as in (S) */
  if (no accesses left)
    set Π_j to 0 and continue

  /* Compute penalty π_{j,k} between c_j and c_k */
  for all other clients c_k
    set π_{j,k} to min(ρ_j, ρ_k) × J                   /* eq. (1) */
    if (c_k is already in penalty phase)
      bound π_{j,k} by a priori penalty β_{j,k} of c_k
      subtract π_{j,k} from β_{j,k}
    add π_{j,k} to Π_j

  bound Π_j by outstanding requests α_j                /* as in (S) */
```

Figure 5.5. Function to update penalties Π_j of all active clients c_j for interval J in the iterated worst-case contention algorithm

6. Quality of Memory Models

We deploy our simplified 802.11n setup as described in Section 7 to assess the quality of our models. The base frequency for all clients is 200 MHz and the performance bottleneck is at the first PE, for which the packet throughput in MBit/s is stated in all figures.

Impact of Access Distributions

In literature, memory accesses issued by processing cores are commonly assumed to be exponentially distributed. We confirm that this assumption is also valid for our system, assessing the sensitivity of system performance to access distributions and simulation resolution in single access mode. As memory arbitration is *Round-Robin*, a distribution's burstiness influences the contention. We compare two corner cases, all bursty access and access with inbetween gaps, with the truly random exponential distribution. The simulation resolution is increased by using additional synchronization points (cf. Section 4.0): At 0%, accesses are averaged over complete tasks. At 100%, up to 60 synchronization points per task account for around 100 cycles and 25 accesses each.

As shown in Fig. 5.6, the considered variants of the two corner cases converge with increasing resolution, respectively. Over 12 considered memory frequencies, the average spread from their mean decreases from 2.4% to 0.4% for bursty and from 1.6% to 0.8% for gap patterns. This indicates that the structure within the variants becomes almost irrelevant in the highest simulation resolution. The exponential pattern remains nearly constant for all resolutions.

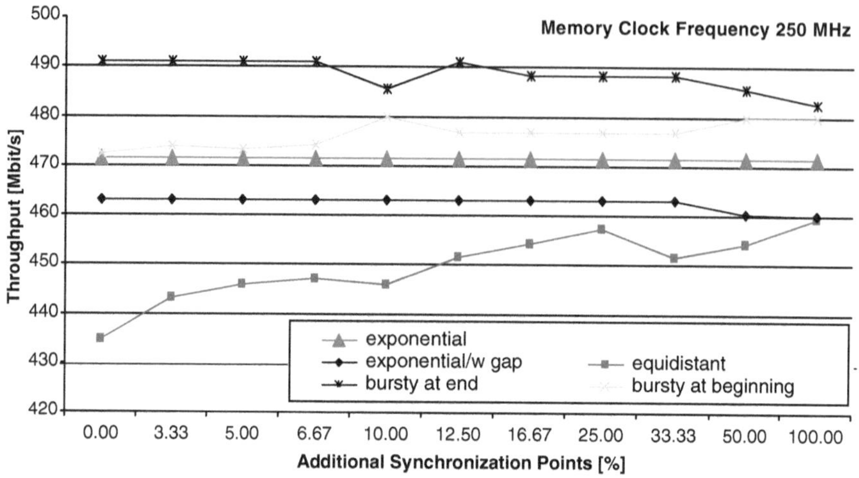

Figure 5.6. Impact of access distribution and synchronization granularity

Over all distributions, the exponential pattern deviates only 0.5% from the mean of the two corner cases. Compared to simulation based on a profiled, cycle-accurate access trace, it deviates only by 1.5%. Thus, the exponential distribution is a good approximation for our system, even at a low simulation resolution.

Comparison of Access Models

A comparison of single and grouped access models in terms of system throughput over memory clock frequency is shown in Fig. 5.7. A set of 12 frequencies is considered that can be derived by a natural fraction of the PEs' clocks. Still, clock synchronization effects limit the throughput at 133.3, 150, and 250 MHz. This is also well reflected statically in the grouped models. For the single access exponential model, which is used as a reference, the penalty caused by memory contention is up to 23% at 50 MHz, drops to 7.5% at 200 MHz, and vanishes with fast memories.

The iterated worst-case algorithm (P) outperforms the simple approach (S) and predicts a throughput 13% below the reference on average. Based on this worst-case contention, the *grouped avg.* approach estimates the actual contention by dividing the penalty by the number of active clients in every interval. It stays within 2.6% of the reference throughput in the range up to 333 MHz, up to four times better than the approach [6] based on [4], which underestimates contention greatly.

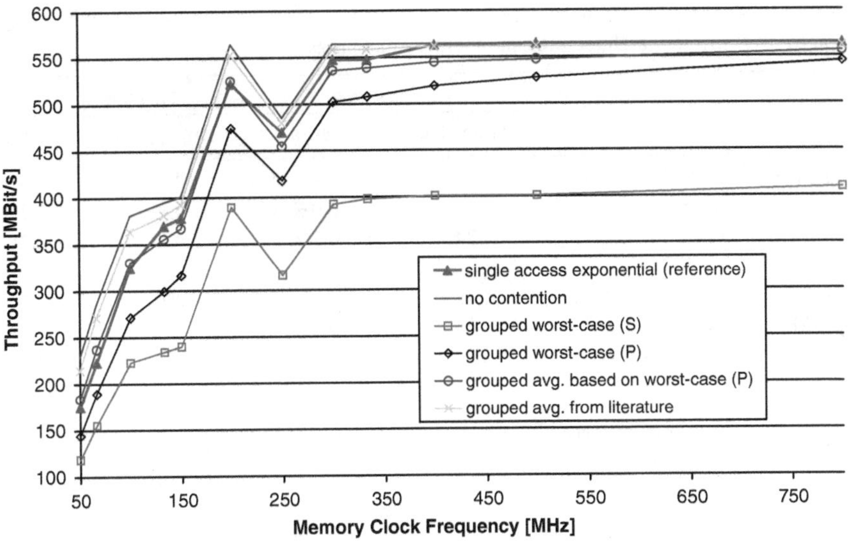

Figure 5.7. Comparison of single and grouped access models

Accuracy of Access Models

The memory models' accuracy when compared to a real system and their sensitivity to changes in system architecture and memory mapping are integral aspects of their qualitity. They are assessed in the following using the preprocessing path of the WLAN application. We anticipate the improved memory mapping of Section 7 for practical reasons, leaving only packets and global data in shared memory. As a reference, a cycle-accurate RTL simulation of an MP-SoC platform [15] is used. This platform together with its memory subsystem is abstracted and accurately represented in SystemClick. Consequently, the simulated execution time of the baseline (single PE, local headers) can be matched almost exactly with the reference.

Representing to two typical steps during exploration, the mapping of a part of the memory is varied (namely packet headers), and the application is partitioned onto a concurrent system environment consisting of 2 PEs and a DMA accelerator. The consequences are different memory access profiles and thus total access latencies due to contention and memory distance. Simulated execution times for our benchmark are summarized in Table 5.1. SystemClick simulation stays within 1% of the reference if the architecture is changed, and within 3.6% if the memory mapping is changed and significantly more shared accesses occur. The single access model outperforms grouped accesses marginally.

Table 5.1. Accuracy of memory access models when compared to cycle-accurate RTL simulation for two different architectures and memory mappings

		Headers local		Headers shared	
		Time[a] [μs]	Accuracy	Time[a] [μs]	Accuracy
Single PE	RTL Simulation	345	–	417	–
	Single Access[b]	345	**0.0%**	429	**2.7%**
	Grouped Access[c]	345	**0.1%**	429	**2.7%**
Two PEs and DMA	RTL Simulation	239	–	306	–
	Single Access[b]	241	**0.8%**	309	**1.2%**
	Grouped Access[c]	241	**1.0%**	317	**3.6%**

[a] Simulated execution time for preprocessing a burst of 15 packets (WLAN application)
[b] SystemClick simulation with exponential distribution
[c] Using the *grouped avg.* model.

Table 5.2. Performance impact in terms of simulated execution time when migrating to a concurrent architecture or allocating more data to shared memory

	Changing architecture		Changing mem. mapping	
	Headers local	Headers shared	1 PE	2PE+DMA
RTL Simulation	−30.9%	−26.7%	20.9%	28.1%
Single Access	−30.3%	−27.8%	24.2%	28.6%
Grouped Access	−30.1%	−26.1%	24.2%	31.3%

More important than absolute accuracy is the sensitivity of the models to these changes. From Table 5.2 it can be observed that both the performance improvement of migrating to the concurrent architecture (up to 31%) as well as the penalty due to allocating more data to shared memory (up to 28%) are reflected accurately in SystemClick. As expected, increasing concurrency is less beneficial for the mapping with more shared accesses due to contention. In summary, this allows quantitative analyses during early design space exploration.

Simulation Performance

Simulation performance depends on access modes and synchronization granularity. As shown in Fig. 5.8 for the five client setup, the grouped access modes outperform the single mode by 50X to 150X, depending on synchronization. This is mostly due to the 330X increase in SystemC events (not shown in figure). The grouped modes triple their runtime with increasing synchronization, also due to event processing.

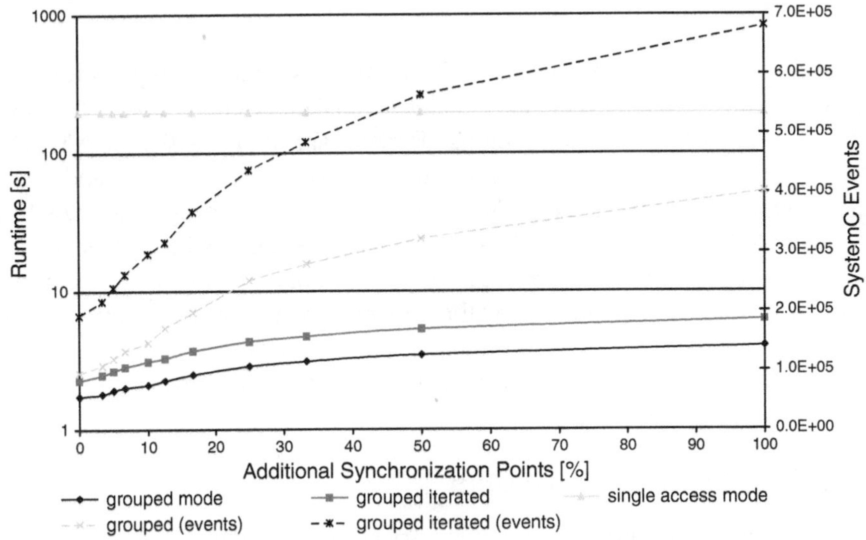

Figure 5.8. Run time (in [s] on the left axis) and SystemC events processed for 10 ms system simulation with increasing number of additional synchronization points

Table 5.3. Simulation speedup of SystemClick over cycle-accurate RTL simulation

	Additional	Simulation speedup[a] [X]		
	Sync. points	Single-acc.	Grouped-acc.	(Incl. setup)
Single PE baseline	100%	2,700	**3,110**	(7,270)
Two PE shared	100%	**970**	840	(2,170)
Two PE shared	0%	1,440	**11,250**	(25,320)

[a]Modelsim 6.1 and open-source SystemC kernel running on same machine.

In addition, the total number of shared memory accesses and the simulated architecture can have a significant impact. Table 5.3 exposes the speedup over RTL simulation for the setting of the preceding section. The single-PE baseline achieves 3,000X speedup, where the advantage of grouped mode is small in the face of fewer shared memory accesses and tight synchronization. With the more complex, concurrent architecture and more memory accesses the speedup is still almost 1,000X. If synchronization is relaxed, grouped access mode achieves a total speedup of 4 orders of magnitude, eight times faster than single access mode. The speedup can be as high as 25k if the setup time for the platform's OS and the application is included.

7. Case Study

We use the transmission path of an 802.11n WLAN reference application [16] as a case study. Processing includes header translation, encryption, rate selection, QoS classification, and queuing. Packets are aggregated into AMSDU frames and further at the PHY layer, requiring complex reordering and block acknowledgment mechanisms. The system function is mapped to two PEs (cf. Section 2) running at 200 MHz. Including three DMAs for packet IO and encryption, five clients access a single shared memory. The impact of increased memory distance is explored as well as memory mappings and optimizations to achieve 450 MBit/s target throughput (3×3 MIMO). For timing-precise performance simulation, a database has been profiled using cycle-accurate instruction set simulation (Mips M4K).

Impact of Memory Subsystem

The communication system runs at memory speed as in Fig. 5.3. A request and response delay of one PE cycle occurs. Arbitration takes one memory cycle and additional read memory latency cycles reflect shared memory distance: A *close* on-chip memory is represented by no extra latency whereas *distant* off-chip SRAM requires three extra cycles.

Mapping all data to shared memory, we observe from Fig. 5.9 that none of the configurations achieves the target. The performance bottleneck are the

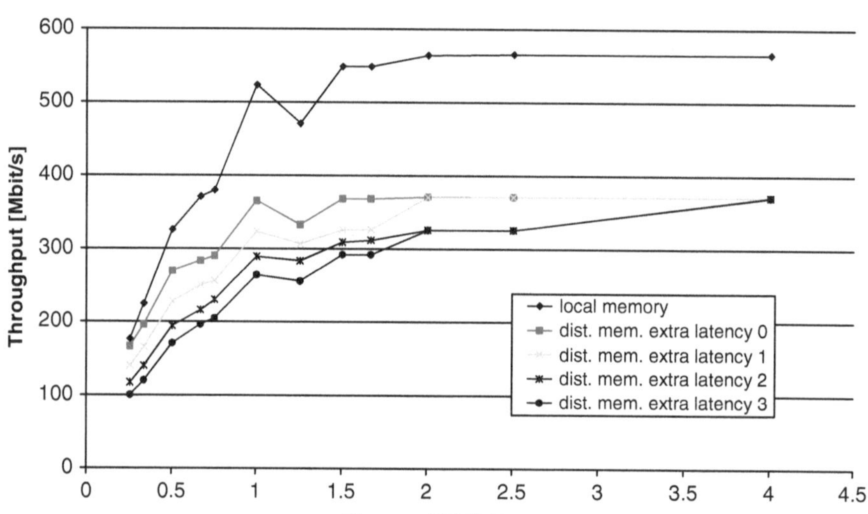

Figure 5.9. Increasing shared memory distance and access contention lowers system performance, shown here over varying memory clock speeds

preprocessing functions being executed on the first PE. Performance is down 30% even when PE and a close memory run at the same speed. Moving to distant memory, the performance drops again by 30%. Increasing the latency further (e.g., to represent SD-RAM) starves the PEs significantly and is not appropriate for frequently accessed data structures in uncached systems.

Very fast memories are unrealistic and their benefit is shown to be limited as part of the total memory latency depends on the PE clock. In addition, the effect of synchronizing the local PE clock with the memory subsystem also decreases performance if no proper fractions are chosen. For example, at 1.25X the PE speed, the penalty is 12% not accounting for additional synchronizing logic that would be required. We thus limit the exploration to memories running 0.25, 0.5 and 1X the PE speed.

Application and Memory Mappings

Mapping part of the 200 kB application memory to closely-coupled private memories mitigates the adverse impact of shared memory latency and contention. But private memories are expensive, need to be allocated for each PE, and limit the flexible use of the platform's memory resources. Thus, we aim to keep these as small as possible.

Mappings of six areas are considered, including stack, local heap and element state, and packet descriptors. A packet buffer sized at 180 kB and global configuration need always be shared. Representing an optimization step of the application itself, we also map packet headers locally. This comes at the expense of more complex SW and extra communication in the final system. Figure 5.10 shows the result of an exhaustive search of all 64 mapping combinations, with Pareto-optimal points achieving 450 Mbit/s being highlighted.

For a close shared memory, the target is met at 200 MHz at a cost of 12 kB of local memory for element configuration, ports, and stack. Up to 31 kB are needed for a memory at 100 MHz. However, with the packet header also kept locally, 13 kB are sufficient. For a distant memory, the speed must be doubled to 200 MHz for the same mapping. At the expense of 32 kB, when only queues are shared, a close memory even at 50 MHz and the 100 MHz distant memory become feasible. At 50 MHz, however, the memory system's capacity limits a sustained throughput.

The penalty of moving from a close to a distant memory at the same speed is between 28% and 42% if all data is shared, and between 6% and 23% if all possible data is mapped locally. The penalty is highest for very slow memories.

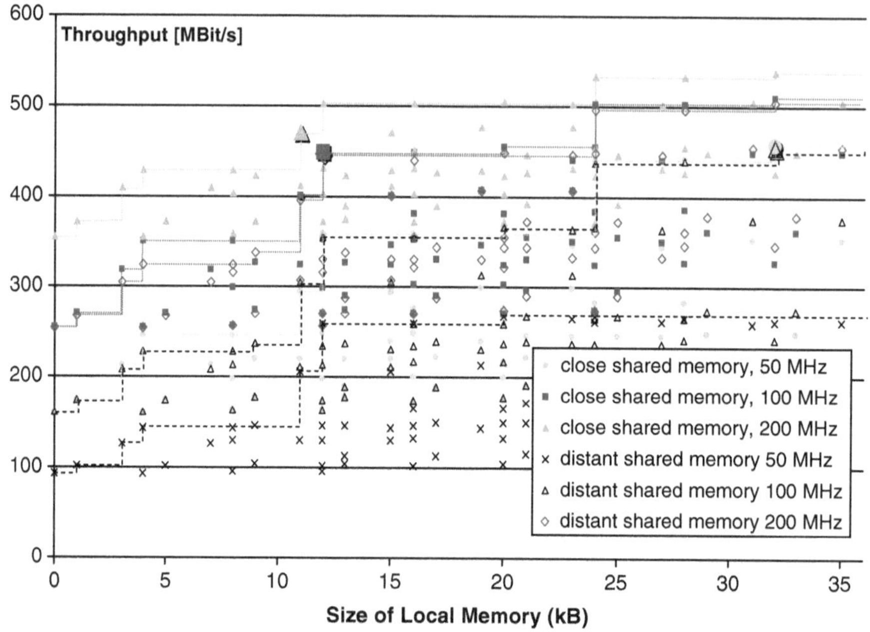

Figure 5.10. Impact of memory mappings reflected by local memory size on system throughput. Pareto fronts are highlighted

Discussion

The memory subsystem can severely impair system throughput. However, this effect can be limited if clocking, memory architecture, and memory mapping are chosen appropriately. Especially:

- The memory subsystem's frequency should be reasonable divider of the PEs' frequency due to synchronization effects (up to 12%). Very fast memory subsystems are unrealistic and not beneficial.

- Allocating only 5% (12 KB) of total memory locally yields a performance gain of 25% and allows to reach the target throughput for a close 200 MHz memory in the case study. For a memory at half the PE speed, 15% of the memory or the evaluated packet header optimization step are needed (6%). Mapping of stack and element state locally is vital.

- Distant off-chip memory for frequently accessed shared data should only be used if it is a design requirement or significantly cheaper. It comes at the cost of decreased performance (up to 42%) and in general requires more local memory (e.g., 12–31 KB for 100 MHz memories), higher memory frequencies (e.g., more than 200 MHz instead of 100 MHz), or a significant increase in PE speed.

- The impact of contention on throughput was up to 20% in some case but dropped, e.g., to <3% for larger local memories.

The Pareto-optimal points of the *grouped avg.* access model were again within few percent of the single access exponential model reference, which did not influence the choice of solutions. The 384 design points can be evaluated in less than 1 hour on a 1.7 GHz laptop with 2 GB memory running an OSCI+Cygwin configuration.

8. Conclusion

We have presented an extension to the SystemClick framework that is able to accurately reflect shared memory penalties due to latency, arbitration, and synchronization with two different memory models. In single access mode, stimulus to abstracted memories or SystemC models can be generated, and the impact of access distributions can be explored.

The improved algorithm for grouped accesses was able to predict performance penalties in our design due to contention up to four times better than a previous approach from literature. In addition, absolute upper and lower bounds for these penalties can be given. The grouped access mode runs up to 150X faster than single access mode due to less simulation events. The tool proved to be both productive and four orders of magnitude faster than RTL simulation allowing to quantitatively evaluate a large number of design points.

For the packet processing application explored, it was shown that by allocating 5% of total memory locally performance can be improved by 25%. The optimization step considered allowed further to half the needed memory speed or to use a further distant memory.

Acknowledgment

This work is partly funded by the EU's FP7 programme under grant no. 213311 (OMEGA).

References

[1] T. Kempf, K. Karuri, S. Wallentowitz, G. Ascheid, R. Leupers, et al., A SW performance estimation framework for early system-level-design using fine-grained instrumentation, in *DATE*, 2006

[2] C. Sauer, M. Gries, H.P. Loeb, SystemClick – a domain-specific framework for early exploration using functional performance models, in *DAC*, 2008

[3] T.F. Tsuei, M.K. Vernon, A multiprocessor bus design model validated by system measurement, in *IEEE Transactions on Parallel Distributed Systems*, 1992

[4] C.-H. Chen, F.-F. Lin, An easy-to-use approach for practical bus-based system design, in *IEEE Transactions on Computers,* 1999

[5] L. Gao, S. Kraemer, et al., An integrated performance estimation approach in a hybrid simulation framework, in *MoBS*, 2008

[6] A. Bobrek, J. Pieper, J. Nelson, J. Paul, D. Thomas, Modeling shared resource contention using a hybrid simulation/analytical approach, in *DATE*, 2004

[7] A. Bobrek, J. Paul, D. Thomas, Shared resource access attributes for high-level contention models, in *DAC*, 2007

[8] S. Schliecker, A. Hamann, R. Racu, R. Ernst, Formal methods for system level performance analysis and optimization, in *Proceedings of the Design Verification Conference*, 2008

[9] S. Künzli, F. Poletti, L. Benini, L. Thiele, Combining simulation and formal methods for system-level performance analysis, in *DATE*, 2006

[10] K. Lahiri, A. Raghunathan, S. Dey, System-level performance analysis for designing on-chip communication architectures, *IEEE Transactions on Computer Aided-Design of Integrated Circuits and Systems*, 2001

[11] O. Ozturk, M. Kandemir, M. Irwin, S. Tosun, Multi-level on-chip memory hierarchy design for embedded chip multiprocessors, in *ICPADS*, 2006

[12] P.R. Panda, F. Catthoor, et al., Data and memory optimization techniques for embedded systems, *ACM TDAES*, 2001

[13] M. Verma, S. Steinke, P. Marwedel, Data partitioning for maximal scratchpad usage, in *ASPDAC*, 2003

[14] A. Gordon-Ross, N. Dutt, et al., Automatic tuning of twolevel caches to embedded applications, in *DATE*, 2004

[15] C. Sauer, M. Gries, S. Dirk, Hard- and software modularity of the NOVA MPSoC platform, in *DATE.* San Jose, CA, USA: EDA Consortium, 2007, pp. 1102–1107

[16] H.-P. Loeb, C. Sauer, A modular reference application for IEEE 802.11n Wireless LAN MACs, in *ICC*, 2009

Chapter 6

ANOTHER TAKE ON FUNCTIONAL SYSTEM-LEVEL DESIGN AND MODELING

Tomasz Toczek, Dominique Houzet, and Stéphane Mancini
961 rue de la Houille Blanche, BP 46 F - 38402 GRENOBLE Cedex, France
{ tomasz.toczek, dominique.houzet, stephane.mancini } @gipsa-lab.inpg.fr

Abstract This chapter exposes a novel methodology suited for efficiently solving problems such as NoC instantiation or memory hierarchy structure determination, common in high-complexity SoCs. The proposed framework is not specific to a given NoC or memory organisation, and supports multiple and mixed abstraction levels and design paradigms. It is based on functional programing techniques such as polymorphism and monadic programming. It has been prototyped in the Haskell language. We show its usefulness on a simple example of estimation of memory accesses impact during the backprojection step of a positron emission tomography (PET) reconstruction algorithm.

Keywords: System level design, Functional programming, Haskell, Tomographic reconstruction

1. Introduction

EDA design has been a field of high and diverse research efforts over the last decade. The growing complexity of electronic designs have stimulated the development of new approaches, such as NoC design and instantiation, high level synthesis, design space exploration, performance estimation and analysis, and so on. Oftentimes, proposed approaches consist in adding a step to the design flow tackling one of the specific problems listed above. This leads to mutual incompatibilities between such approaches which could be avoided with a sounder design.

Using functional programming for hardware design tends to be seen as a good idea [12, 14]. Typed functional languages are naturally apt to handle the data representations likely to emerge when describing hardware: pattern matching makes graph transformations easy to express, and lambda-calculus

D. Borrione (ed.), *Advances in Design Methods from Modeling Languages for Embedded Systems and SoC's*, Lecture Notes in Electrical Engineering 63,
DOI 10.1007/978-90-481-9304-2_6, © Springer Science+Business Media B.V. 2010

permits the design of efficient combinator-based hardware instantiation libraries. Bearing this in mind, we try and propose an API with as clear and modular a design as possible. This way, with a single added step in the design flow, a significant number of unrelated optimizations or design techniques can be expressed and used.

2. State of the Art

Several attempts to use functional programming for hardware description have already been made. One of the most well known is Bluespec System Verilog (BSV), a commercial RTL level design tool [2]. It is a good example of what equational reasoning can bring to HDLs. Hardware descriptions in BSV tend to be more concise and less error prone than, say, their VHDL or Verilog counterparts. This is mainly due to the block instantiation facilities, which permit increased code reuse and automated glue logic generation.

Several academic projects have also tackled the issue through a broad range of approaches. Lava [1] and ForSyDe [10], which also use Haskell's polymorphism as a foundation, are among the ones which relate the most to the methodology described in the chapter. Just like BSV, Lava expresses hardware at the RTL level. Its main feature is to allow the same polymorphic construct to be used both for synthesis, simulation, verification, etc., depending on which actual type it is eventually bound to. In a way, one might say our approach merely tries to generalize this to a maximum extent. ForSyDe focuses on transformational design refinement from high level design specifications. Since communication and computation are separated, the ForSyDe methodology can be adapted to several models of computations (MoCs). It should be noted that mixing MoCs is discouraged [10], however. From the user's standpoint, ForSyDe's approach, just like Lava's and unlike ours, is an essentially combinator-based – as opposed to monadic – but makes use of higher order functions nonetheless.

Another noteworthy tool is SML-sys [6], which emphasises on providing multiple MoCs for simulation purposes in Standard ML. Since Standard ML lacks a class system such as Haskell's, SML-sys is less reliant on polymorphic tricks than other functional approaches might be; for instance, extending it for other purposes than simulation is not straightforward. Its strength on the contrary resides in its rich library of MoCs.

Ωmega [13], a dependently typed language, has been shown as by its creator to be suited for hardware description and verification using the Curry-Howard isomorphism [12]. Even though the examples given are expressed at the gate level, they clearly make their point at showing the utility of dependent types for property verification. While Ωmega is still experimental, given its syntactic similarities with Haskell, we may hope to see one day a properly specified

Haskell variant with both classes and dependent types, and theorem libraries like Agda's [8]. This makes Haskell-based approaches all the more relevant.

Of course, classical approaches to the system-level design problem exist, a lot of which offer a higher level of maturity. This is for instance the case with Ptolemy II [5], a set of software components in Java for actor-oriented architecture design supporting a very broad range of MoCs. In a way, it can be argued that the different SystemC 2.0 interfaces, allowing for levels of modeling ranging from transactional to cycle-accurate, also relax the model of computation imposed by the language. What is more, there has been several projects aiming at providing SystemC with fully-fledged multi-MoC modeling capabilities [4, 9]. While often quite efficient, those methods do not offset the specific gains, such as a higher expressivity and formalism, which can be brought by purely functional approaches.

3. Overview

Engine/MoC Dichotomy

We make an essential distinction between what we call *engines* and *MoCs*. For us, a *MoC* is simply a class specifying a limited set of primitives and communication mechanisms allowing the user to express some part of the design. Examples would include synchronous RTL, untimed imperative style specification, KPN-like descriptions, and so on. Since a MoC is a simple Haskell class *declaration* (and not a specific instance of that class), writing a new one is rather easy but not really useful by itself, i.e. without an engine. A MoC may of course have associated libraries (containing additional primitives, expressed using the more fundamental ones).

MoC instances are what we call *engines*. Engines can have different capabilities, such as simulation, intermediary HDL code production, and so on. A MoC may have several engines, and it is worth noticing that it may also make sense for a given engine to instantiate several MoCs; the degree of specialisation of engines may vary. As an example, supposing one defined the synchronous RTL and untimed MoCs, we could associate to them a common simulation engine, a VHDL generation engine for the synchronous RTL MoC only, and so on. Since the engines are where things really get done, they tend to be more complex to implement than simple MoC specifications.

For the time being, we have a working implementation of a versatile RTL level engine, as well as a simple instrumentation engine. The latter will be presented in Section 4. Critical parts of an engine may be implemented in C and given Haskell bindings.

Please note that the user is incited to describe his hardware polymorphically, using a MoC and not an engine. The binding of the design to an engine should be done in the main function of a Haskell program doing some sort of operation on the design: performance estimation, HDL generation for synthesis, and so on. Design specification and design utilisation should be clearly distinct. This aspect is similar to the workings of Lava.

Data Exchange Between MoCs

In order to make cooperation between MoCs easier, each MoC (as a class) should take at least two parameters: one corresponding to a type representing the design itself, or the relevant information extracted thereof, and several types representing the kinds of objects handled by the MoC – what we call associated objects. The "meaning" of the objects associated to a MoC generally depends on its abstraction level, how time is handled, etc. For instance, for MoCs representing synchronous circuits, one associated object type could be a "clock domain" type. It is then particularly convenient to use type families, as implemented for instance in GHC [11], to define entities such as, for instance, registers (with their clock domain). Examples will follow.

Using multi-parameter classes makes the approach non-compliant with Haskell 98, but we rely on a lot of non-standard features of GHC anyway, most of which being meant to be integrated into Haskell'.

MoC Design

We have respected several rather simple guidelines for the MoC classes in our implementation. Namely:

- We implemented MoCs as monads, resulting for the end-user in a design specification style often close to imperative programming. Since most users are accustomed to such programming styles, it increases the overall ease of use. Monadic programming is not a strict requirement for our propositions to make sense, but provides significant advantages (see next item as well as Section 3.0).
- A MoC class should be both:
  - Minimalist from the point of view of the described primitives. As already mentioned, one can safely create libraries of more complex constructs from those primitives, and keeping the MoC class itself simple makes the implementation of engines easier.
  - As rich as possible in terms of compatibility with other MoCs and other helper components. For instance, we define an *Explorable* monad class, with primitives for expressing alternative implementation of some parts of the design (see Section 3.0). Each of our MoCs classes is an explorable monad. One might think the most elegant course of action is to let the user decide explicitly on a case

by case basis which classes a design matches, but this is not really conveniently feasible in practice. Indeed, we often need to embed MoCs one into another, the embedded sub-design "inheriting" some of the features of the encompassing design (such as explorability). This leads to higher rank polymorphic embedding functions, with signatures such as *embed :: (forall a. EmbeddedMoC a => a) -> (forall a. EncompassingMoC a => a)*; in other words, we transform a polymorphic expression into another. If the encompassing design has the type *forall a. (EncompassingMoC a, AdditionalMoCFeature a) => a*, there is no way to ensure that the embedded MoC is an instance of *AdditionalMoCFeature* as well. The *embed* function could be specialised to make *AdditionalMoCFeature* a requirement for both the encompassing and embedded MoCs, but then could not be used when neither of the MoCs have that feature. In the end, the simplest way to deal with this problem is to make *AdditionalMoCFeature* a requirement of both MoCs, even if it goes unused sometimes.

Engine Design

Our guidelines for engine design where:

- Making it easy to embed an engine into another, so as to share common functionality between engines.
- While a MoC should contain a reduced set of primitives, an engine may (and should) come with a rich set of functions allowing to make efficient usage of the data generated from a design.

In order to meet these goals, we defined an explorable state monad data type. Any engine implemented using the *ExpState* data structure is simply defined by a state representation and a set of modifications of it. This greatly simplifies its design. The explorable state monad is a simple variant of the regular state monad, instantiating the above mentioned *Explorable* class for the expression of alternatives. This leads to an implementation such as shown in Listing 6.1.

The *ExpInfo* class denotes helper types used for exploration purposes of *ExpState* monads. As can be inferred from the *ExpState* definition, each bind updates not only the state, but also some instance of *ExpInfo*. The *ExpInfo* object can be seen as the useful information about the exploration process in general, and will depend on the kind of exploration algorithm used. Typical cases include information about the alternatives which were chosen to update the state (for methods such as exhaustive exploration, hill climbing,...), and lists of pseudo-random numbers when using stochastic exploration methods.

```
class Monad m => Explorable m where
    alternatives :: [m a] -> m a

idle = return ()

class ExpInfo info where
    eiCall ::
        (s -> (a, s)) -> info -> s -> (info, a, s)
    eiAlternatives ::
        [ExpState s a] -> info -> s -> (info, a, s)

data ExpState s a = ExpState {
    runExpState :: ExpInfo info =>
        info -> s -> (info, a, s)
}

instance Monad (ExpState s) where
    return x = ExpState $ \info s -> (info, x, s)
    m >>= k =
        let f info s =
            let (info', a, s') =
                (runExpState m) info s
            in (runExpState (k a)) info' s'
        in ExpState $ f

eiModify f = let g s = ((), f s) in eiCall g
call f = ExpState (eiCall f)
modify f = ExpState (eiModify f)
oneOf xs = alternatives (map return xs)
```

Listing 6.1. The explorable monad class, and the explorable state monad data type

4. Examples

Instrumentation

Often, a design needs to be instrumented in order to extract relevant information. In this section, we describe a MoC allowing such instrumentation. Used on its own, it can be convenient to assess some properties of high level models of designs – we show this in the next section. It can (and should!) also be used as a dependency for other MoC, in order to allow the instrumentation of fully expressed designs instead of simple models.

As can be seen in Listing 6.2, the instrumentation MoC offers two main concepts: files and accumulators. The *open* function selects the current file, and *append* function concatenates a string to it. The accumulators are, on the other hand, integer variables which may be used for various purposes. A new accumulator is created with the *namedAccu* function. Its value can be read with the help of *accuVal*. For convenience, we define C-like incrementation and decrementation operators on accumulators, as well as a way to specify the

```
type family Accu en :: *

class Explorable em
    => Instrumented en em | em -> en where

    -- Mendaroty (accumulator handling)
    namedAccu :: String -> em (Accu en)
    (#=) :: Accu en -> Integer -> em ()
    accuVal :: Accu en -> em Integer

    -- Mendatory (file handling)
    open :: String -> em ()
    append :: String -> em ()

    -- With default implementations
    accu :: em (Accu en)
    (+=) :: Accu en -> Integer -> em ()
    (-=) :: Accu en -> Integer -> em ()
    setPrettyPrinter ::
        Accu en -> (Integer -> String) -> em ()

    accu = namedAccu ''<anonymous accu>''
    x += y = do tmp <- accuVal x; x #= (tmp + y)
    x -= y = do tmp <- accuVal x; x #= (tmp - y)
    setPrettyPrinter accu printer = idle

while :: Instrumented en em =>
    em Bool -> em () -> em ()
whilst :: Instrumented en em =>
    Accu en -> (Integer -> Bool) -> em () -> em ()
while cond block = do
    continue <- cond
    if continue
        then do block; while cond block
        else idle
whilst x cond' block =
    let cond = do
        tmp <- accuVal x
        return (cond' tmp)
    in while cond block
```

Listing 6.2. The instrumentation MoC

string representation of a given accumulator. It is up to the engine to use the pretty printing function provided by the user. An accumulator is represented by the *Accu en* type, where *en* is the (only) object associated to the instrumented engine *em*.

We also define a simple instance of *Instrumented*, called *StdInstrumented*. In addition to implementing the primitives required by the MoC, the engine has additional functions allowing it to produce useful tables from the harvested data.

Evaluation of Memory Access Costs for Tomographic Reconstruction

As a real world example, we chose to study the memory accesses patterns during the backprojection step of a PET reconstruction algorithm, as given in [3]. The principle of tomographic backprojection is to compute a voxelized (that is, using voxels, i.e. "volume pixels") representation of a volume from its projection of some kind. The projection data itself is provided as a multi-dimensional array called a sinogram, and generally obtained using an MRI scanner.

Cache Memory Models. The study relies on timeless models of cache memories implemented within the instrumentation MoC, as shown in Listing 6.3. We limited ourselves to set-associative 1D caches, created by the *cache* function and read with the *cacheRead* function. On a miss, the new data is put into the slot which has not been read for the longest time – more content management policies could be implemented. The significant performance related parameters of the cache are the cache line size, the number of slots, and the set size. For convenience, *DirectMapped* and *FullyAssociative* can be used as aliases for 1-way associative and n-way associative, respectively, n being the number of slots of the cache.

Since we need to handle 3D data, we implemented convenience functions to acces nD arrays within our 1D cache. The *cacheND* and *cacheNDRead* functions can be used to that effect, completely abstracting the underlying set associative cache. The user needs to provide an addressing scheme for the array at the creation of the cache, as well as the array dimensions, and all the above-mentioned parameters ruling the behaviour of the set-associative cache used in practice. For our tests, we used three addressing schemes, as shown on Fig. 6.1: the classical "scanline" order – or rather, its n-dimensional generalisation (*addrNDZLinear*), and two variants of the z-order curve [7], one obtained be interleaving the least significant bits of the coordinates first (*addrNDZOrder*), and the other by interleaving the most significant first (*addrNDZOrder2*). This does not make a difference if the array has the same size along all the dimensions.

As the values read from memory do not impact the data path in our application, the cache models do not contain the actual data, and the reading function does not "return" anything – it only updates the internal state information of the cache. A variant with the data actually stored could easily be imagined, and implemented.

User Specified Design. The design itself, as should be provided by the user, is expressed in the Listing 6.4. The parameters which are allowed to vary are expressed using the *oneOf* helper function, which relies on the *alternatives* primitive of the explorable monads. In our case, we allow several values for

```
data CacheAssociativity
    = FullyAssociative
    | DirectMapped
    | SetAssociative Integer -- slots per set

data Cache = Cache
    { cacheName :: String
    , cacheAssociativity :: CacheAssociativity
    , cacheLineSize :: Integer
    , cacheNSlots :: Integer
    }

data CacheId en = -- (...)

cache :: Instrumented en em =>
    Cache -> em (CacheId en)
cacheRead :: Instrumented en em =>
    CacheId en -> Integer -> Integer -> em ()

data CacheND = DumbCacheND
    { cacheNDName :: String
    , cacheNDBytesPerElement :: Integer
    , cacheNDDims :: [Integer]
    , cacheNDAssociativity :: CacheAssociativity
    , cacheNDLineSize :: Integer
    , cacheNDNSlots :: Integer
    , cacheNDAddr :: CacheND -> [Integer] -> Integer
    }

data CacheNDId en = -- (...)

cacheND :: Instrumented en em =>
    CacheND -> em (CacheNDId en)
cacheNDRead :: Instrumented en em =>
    CacheNDId en -> [Integer] -> em ()

addrNDLinear :: CacheND -> [Integer] -> Integer
addrNDZOrder :: CacheND -> [Integer] -> Integer
addrNDZOrder2 :: CacheND -> [Integer] -> Integer
```

Listing 6.3. Cache model interface

the cache line size, the number of slots, the set size, and the sinogram adressing scheme. Dependancies between such parameters are allowed; for example, a set cannot have more slots than the cache has. The exploration functions are meant to be robust against such cases. Instrumentation of values which do not vary is eased by the *instrument* function, which automatically creates the corresponding accumulators. The main loop itself consists in iterating on ψ, and performing memory accesses accordingly, using the u and v functions to determine the relevant point of the sinogram. To keep things simple, only eight adjacent voxels ($2 \times 2 \times 2$) have their values computed. Gac et al. [3] (from which we borrow most of our variable names) have indeed shown that performing the

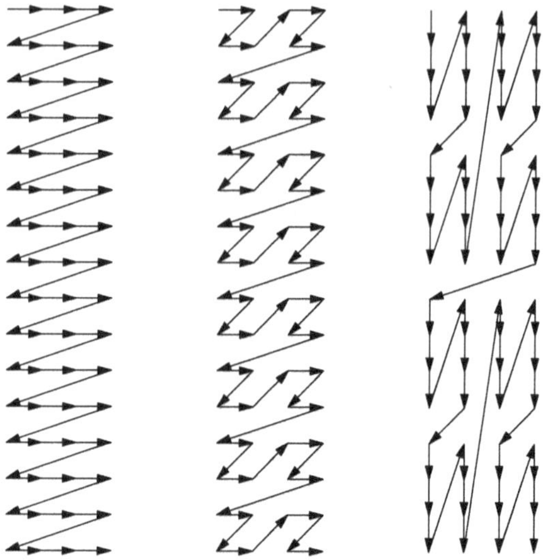

Figure 6.1. From left to right: index order in a 4×16 array, using the *addrNDLinear*, *addrNDZOrder* and *addrNDZOrder2* indexing schemes

backprojection iterating on blocks of voxels instead of single voxels significantly increases memory access locality. The position of our example block in the volume has been arbitrarily chosen.

The resulting spatial distribution of reads is shown on the Fig. 6.2. The reads basically follow a three-dimensional sine wave. The criterion we chose to measure the memory access performance is the number of burst accesses to external memory needed to retrieve all the information requested. The results are presented on the Figs. 6.3, 6.4 and 6.5. In order to obtain these results, we tested all the possible parameters combinations. Each cell of the tables gives the performance range obtained by varying the parameters which have not been fixed.

Performance Discussion. The conclusion which can be drawn from this study is that one of the main performance criterion in such an application is the addressing scheme. In our case, since we used a sinogram with a highly skewed aspect ratio ($256 \times 256 \times 32$), the variant of the z-curve obtained by interleaving the most significant bits first seems to work the best. We can also see that, when using the classic scanline ordering, the performance increases gradually when adding supplementary slots to the cache, whereas it peaks very quickly (little gain beyond four slots) with the z-order based addressing schemes – at least in the best cases. The impact of the associativity of the cache can be seen on the Fig. 6.3. A highly associative cache helps greatly especially in the worst cases, typically when the addressing scheme fails to preserve locality. If

```
tomo :: Instrumented en em => em ()
tomo =
    let
        (psiMax, uMax, vMax) = (256, 256, 32)
        dims = [psiMax, uMax, vMax]
        -- (x, y, z) are meant to be in [-1.0, 1.0]^3
        r@(xo, yo, zo) = (0.2, 0.3, 0.1)
        (dx, dy, dz) = (2.0/64.0, 2.0/64.0, 2.0/64.0)
        (bx, by, bz) = (2, 2, 2) -- block size
        delta = 1.0
        u psi r@(x, y, z) =
            let rv = (x * cos psi + y * sin psi + 1.0)
            in floor (rv * 0.5 * fromInteger uMax)
        v psi delta r@(x, y, z) =
            let a = (x * sin psi - y * cos psi)
                rv = delta / (2.0*1.0) * a + z + 2.0
            in floor (rv * 0.25 * fromInteger vMax)
        psiStep = 2.0 * 3.14159/(fromIntegral psiMax)
        getCoords it r =
            let psi = (fromIntegral it) * psiStep
            in [it, u psi r, v psi delta r]
    in do
        lineSize <- oneOf [128, 256]
        nSlots <- oneOf [1, 2, 4, 8, 16]
        nAs <- oneOf (possibleAs nSlots)
        order <- oneOf [0, 1, 2]
        addr <- case order of
            0 -> return addrNDLinear
            1 -> return addrNDZOrder
            2 -> return addrNDZOrder2
        cache <- cacheND (DumbCacheND ''cache'' 4 dims
            (SetAssociative nAs) lineSize nSlots addr)
        instrument [
            (nAs, ''associativity''),
            (nSlots, ''# slots''),
            (lineSize, ''line size''),
            (order, ''adressing'')
        ]

        currentPsi <- namedAccu ''current psi''
        whilst currentPsi (< psiMax)
          ( do
            it <- accuVal currentPsi
            [0 .. (bx*by*bz - 1)] `for` \i ->
                let
                    xi = i `rem` bx
                    yi = (i `quot` bx) `rem` by
                    zi = i `quot` (bx*by)
                    x = xo + dx * fromIntegral xi
                    y = yo + dy * fromIntegral yi
                    z = zo + dz * fromIntegral zi
                    r = (x, y, z)
                in cacheNDRead cache (getCoords it r)
            currentPsi += 1
          )
```

Listing 6.4. PET backprojection loop model

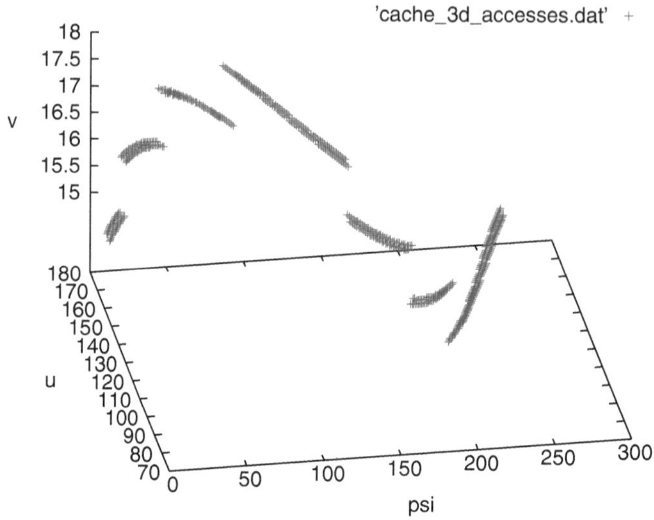

Figure 6.2. The spatial memory access patterns for the determination of a $2 \times 2 \times 2$ block of voxels during the backprojection

Associativity	# slots				
	1	2	4	8	16
Direct mapped	1,473 –1,922	574 –1,922	574 –1,922	380 –1,922	152 –1,922
2-way		646 –1,754	163 –1,754	163 –1,754	152 –1,754
4-way			174 –847	152 –847	152 –847
8-way				152 –447	152 –447
16-way					152 –371

Figure 6.3. Burst accesses to RAM, depending on the number of cache slots and the size of sets

# Slots	Adressing		
	Linear	Z-order	Z-order (MSB first)
1	1,922	1,473–1,779	1,494–1,502
2	1,752–1,922	646–1,623	574–1,502
4	839–1,922	231–1,606	163–725
8	411–1,922	215–1,606	152–613
16	326–1,922	215–380	152–449

Figure 6.4. Burst accesses to RAM, depending on the number of cache slots and the adressing scheme

# Slots	Line size	
	64	128
1	1,502–1,922	1,473–1,922
2	725–1,922	574–1,922
4	243–1,922	163–1,922
8	227–1,922	152–1,922
16	227–1,922	152–1,784

Figure 6.5. Burst accesses to RAM, depending on the number of cache slots and cache line size

we were to make an architectural choice based on this data, it seems a two-way associative cache with four slots, 128 byte burst accesses and the MSB interleaving variant of z-order addressing is a good compromise. Because of a slight inadequacy of our slot replacement policy, it even performs a bit better that its fully associative counterpart in this very case.

5. Conclusion

We showed several programming principles allowing to construct a flexible design framework, easily extended with new models, new MoCs, and new ways to extract relevant data. We have shown on a real-life example a small subset of the those aspects. The proposed approach is sound, and does not seem clumsy despite the fact that Haskell is not genuinely a hardware design or modelling language.

A lot of potential gains of the proposed method do not show, however, on the case study given. Our method is meant to act especially well on big and heterogeneous designs, whose expression is hindered by their complexity using conventional approaches. MoC mixing and communication play a major role in such cases, while being especially easy to express in our approach. Furthermore, handling large design spaces is much less of a problem as long as suitable exploration methods are used – which we allow for. Future work will include specifying a full-scale design using our methodology, to assess those features.

References

[1] P. Bjesse, K. Claessen, M. Sheeran, S. Singh, Lava: hardware design in haskell, In *ICFP '98: Proceedings of the Third ACM SIGPLAN International Conference on Functional Programming* (ACM Press, New York, NY, 1998), pp. 174–184

[2] Bluespec, 2000, `http://www.bluespec.com/`

[3] N. Gac, S. Mancini, M. Desvignes, D. Houzet High speed 3d tomography on CPU, GPU and FPGA. EURASIP J. Embed. Syst. Special issue: design and architectures for signal image processing, vol. 2008, Article ID 930250, Hindawi Publishing Corporation

[4] F. Herrera, E. Villar, A framework for heterogeneous specification and design of electronic embedded systems in systemC. ACM Trans. Des. Autom. Electron. Syst. **12**(3), 1–31 (2007)

[5] C. Hylands, E. Lee, J. Liu, X. Liu, S. Neuendorffer, Y. Xiong, Y. Zhao, H. Zheng, Overview of the ptolemy project. Technical Report UCB/ERL M03/25, EECS Department, University of California, Berkeley, 2003

[6] D. Mathaikutty, H. Patel, S. Shukla, A. Jantsch, Sml-sys: a functional framework with multiple models of computation for modeling heterogeneous system. Design Automat. Embed. Syst. **12**, 1–30 (2008)

[7] G.M. Morton, A computer oriented geodetic data base and a new technique in file sequencing. Technical report, 1966

[8] S.-C. Mu, H.-S. Ko, P. Jansson, Algebra of programming using dependent types, in *MPC '08: Proceedings of the 9th International Conference on Mathematics of Program Construction* (Springer, Berlin, 2008), pp. 268–283

[9] H.D. Patel, S.K. Shukla, R.A. Bergamaschi, Heterogeneous behavioral hierarchy extensions for systemC. IEEE Trans. CAD Integr. Circ. Syst. **26**(4), 765–780 (2007)

[10] I. Sander, A. Jantsch, System modeling and transformational design refinement in ForSyDe. IEEE Trans. Comput. Aided Design Integr. Circ. Syst. **23**(1), 17–32 (2004)

[11] T. Schrijvers, S. Peyton Jones, M. Chakravarty, M. Sulzmann, Type checking with open type functions, 2008

[12] T. Sheard, Types and hardware description languages, in *Hardware Design and Functional Languages, A Satellite Event of ETAPS*, volume 5161, ed. by A. Martin, C. Seger, M. Sheeran, Mary, 2007, Available from: `http://www.cs.pdx.edu/~sheard/`

[13] T. Sheard, N. Linger, Programming in omega, in *CEFP*, volume 5161 of *Lecture Notes in Computer Science*, ed. by Z. Horváth, R. Plasmeijer, A. Soós, V. Zsók (Springer Berlin/Heidelberg, 2007), pp. 158–227, See "http://www.springerlink.com/content/t27647102q527236/" for additional information

[14] M. Sheeran, Hardware design and functional programming: a perfect match. J.UCS. **11**(7), 1135–1158 (2005)

Chapter 7

DESIGN AUTOMATION MODEL FOR APPLICATION-SPECIFIC PROCESSORS ON RECONFIGURABLE FABRIC

Arda Yurdakul, Roza Ghamari, Bayram Kurumahmut, and Gokhan Kabukcu

Computer Engineering Department, Bogazici University, Bebek 34342, Istanbul, Turkey

{yurdakul, roza.ghamari} @boun.edu.tr

Abstract The process of embedded system design on reconfigurable architectures needs smart solutions to reduce development life-cycle and to use resources efficiently at run-time. Current solutions are insufficient to enable the embedded system designer to reflect the flexibility that a reconfigurable architecture can offer. Some of the basic problems are lack of flexible operator definitions, very detailed hardware abstraction procedures, a few or no constraints for tasks or loops at the high-level of design abstraction. In this work, we propose a new model for automated design of application-specific processors in run-time reconfigurable architectures, solving the aforementioned inefficiency problems. Based on the proposal, a design language, a framework and a compiler have also been developed.

Keywords: Embedded system design, Run-time reconfigurable architecture, Hardware software co-design, Control data flow graph

1. Introduction

In the embedded systems arena, there are hundreds of microcontrollers from different manufacturers. However, it is usually impossible to find a microcontroller that exactly meets all requirements of an embedded system. As a result customizable processors [8, 23] have started to appear. On the other hand, reconfigurable devices have a significant place in embedded design domain. Consequently, soft-processors [1], polymorphic processors [12, 24, 26] are proposed. These processors can be tailored to meet the exact requirements of an embedded system. Some of them extends the instruction set while preserving the basic instruction set [1, 8, 23], while newer ones modify the instruction set

D. Borrione (ed.), *Advances in Design Methods from Modeling Languages for Embedded Systems and SoC's*, Lecture Notes in Electrical Engineering 63,
DOI 10.1007/978-90-481-9304-2_7,

by removing unused instructions and/or introducing new instructions [24, 26] or design the complete application-specific instruction set [12]. Most of them support the customizations to be done prior to execution while the newer ones support run-time customization of the instruction set.

There are also design entry tools [8, 20, 22, 27]. The following discussion can be carried out for those tools

- The design tools require detailed hardware definition for the new instructions. Some tools invoke them with function calls which makes embedded programming a confusing issue. ADRES [12] uses a 32-bit three-input/output functional template to define all operations since ADRES is a coarse grain array. As a result, the resources of ADRES are underutilized for simple operations.
- They use programming languages (like C, C++) as they are or extend them. Usually retargetting of the compiler has to be done manually. In some tools, the designer designs the instructions and determines very low details like opcode, operand fields. DRESC provides a parametric template for instruction generation since the underlying architecture, ADRES, is fixed. However, in all of these tools the instruction set is fixed by the user. DRESC allows architecture exploration but instruction generation is not mentioned.
- Retargettable compilers map the built-in operations of the language to the built-in operators of the underlying processor which is assumed to be static during runtime. For example, if the underlying processor is a scalar one, then "+" operation of C will be mapped to the ALU that makes scalar addition. For a vector addition, a new instruction has to be generated and this must be invoked via a function call. When the underlying processor is a runtime reconfigurable one, these tools cannot benefit from this utility of the device. For example, the designer cannot set "+" as a scalar adder in one loop and as a vector adder in another loop.
- They do not enable the user to define timing and resource constraints. If the design does not meet a constraint, then the overall design process has to be revised. Today, design tools for programmable processors [7] support code optimization according to user constraints. However, the tools for reconfigurable ones do not have such an option.

Based on these observations, we propose that a new model and a language must be used for embedded systems on reconfigurable systems. The issues that necessitates a new language are explained in the following section. Then we explain our model, RH(+), which is general enough to handle the run-time reconfigurable processors as well as other types of customizable soft processors and hard processors (Section 3). Based on this model, we developed a new language, LRH(+) which allows the embedded system to be designed with a single language, a framework, FRH(+) which satisfies the basics set by our model, and a compiler which retargets itself automatically by using the templates generated by the instruction set generation tool (Section 4). A design example is presented in Section 5. In the last section, we explain the current status of our project and summarize our future work.

2. Languages for Customizable Processors

Traditionally, programming languages, like C, C++, Java are selected for developing embedded applications. They are also extended to support hardware elements of a system. SpecC, SystemC, Handel-C and Impulse-C are some of the extended-C languages [15, 21]. An embedded design has to be specified to exploit the properties of the target architectures. Today, the most popular extended-programming-language for hardware design is SystemC [15]. There are SystemC-based approaches use a prototyping tool that can estimate area, execution time and reconfiguration overhead [14]. Moreover reconfiguration can be automatically triggered. However, the user-defined system specification constraints like resource and timing constraints, in SystemC [10]. Besides, it does not allow processor customization [19]. Another C-variant language, which is includes design of both hardware and software portions, is SpecC [21]. However, like SystemC, SpecC does not have the features like mapping, estimation and system constraint definition. Impulse-C and Handel-C use C-to-gates tools for architecture mapping to FPGAs. Although they are capable of generating hardware using some units automatically, the software code needs optimizations after compilation [2]. Some low-level timing optimizations are available in Handel-C; however, there is no optimization feature in Impulse-C [2]. Moreover, the operations are predefined in Handel-C; therefore, the reconfigurable system designer encounters with limitations during application development. Extension of programming languages requires modifications in standard compilers in order to be able to separate the hardware and the software.

The modeling languages have also been proposed for modeling embedded systems. SDL and UML [5], are some of the well-known examples. However, their cores are inappropriate to use them as solutions for automatic synthesis. Thus, the community has proposed extensions to adapt them to the applications on reconfigurable hardware xUML [18], xtUML [13]. GASPARD is a platform that introduces synthesis and reconfiguration to a UML-based model MARTE [17] but it cannot be used to generate custom reconfigurable processors. Hence the modeling languages must only be used as representation languages to express the design visually.

In the literature, extending the standard tools is called as *retargetting*. To retarget existing tool suite, architecture description languages (ADL) are used. However, they need low-level details which are irrelevant for software development level. For example MIMOLA models target hardware in a similar manner to VHDL. Thus, it explicitly [6] requires definition of hardware modules with their behavioral algorithm and detailed interconnection scheme between modules. MIMOLA has problems with complex instructions because it is difficult to extract complex instructions from MIMOLA descriptions. As a result, the quality of the produced code is low [9]. nML is another example of formalism which generates description based on both structural and behavioral model, and is applicable for *retargetting* code generation [6]. The main disadvantage of nML is that it requires low-level details to define target architecture. These are behavior, assembly language mnemonic, and binary code of the instructions which are constructed by using predefined operators. LISA [27] is an enhanced ADL compared to nML. However, it has the disadvantage that it requires behavioral description of the extended instructions as nML. In both nML and LISA, new operators (instructions) can be created in two ways. Firstly, predefined operators (+, –, *, etc.) are used to create new operators. Secondly, these created operators can be used to create more complex operators. In an application, both types of operators for the same functionality can be created. Software designer can choose the one providing more efficient solution. However, if software designer uses nML or LISA, he will have a huge work overhead while switching between these two types of implementation. The basic reason is that nML and LISA force the user to give different operator names for each type of implementation. Thus, designer must scan all lines in software, find each references to the operator, and replace it with new name. DRESC uses a parametric template which can be used by the designer to decide on the instructions. Instruction level parallelism is automatically established by DRESC.

3. RH(+) Model

To solve the above-mentioned problems of current design languages, we propose a model, RH(+), for embedded system design on run-time reconfigurable architectures like FPGAs that support dynamic reconfiguration [25]. RH(+) stands for hardware/software co-design on reconfigurable hardware. Based on this model, we propose a new language, LRH(+) which utilizes run-time processor customization and application development in a single language.

Abstraction of Low-Level Details for Instruction Set

For embedded system on reconfigurable hardware, we believe that users should not deal with the details about instruction due to the flexibility of the target hardware. In RH(+), smallest instruction corresponds to an operator. For each operator, it must be sufficient for the designer to enter the number of inputs and outputs of the operator. For example, assume two instructions: A and B. Instruction A might have two inputs, and one output (e.g. scalar addition) and instruction B might have eight inputs, and four outputs (e.g. matrix multiplication of two 4-by-4 matrices). Operators can be combined to generate complex instructions. Complex instruction generation can be done either by the user or left to the design automation tool that generates the optimum instruction set for the application and its imposed constraints.

Flexible Operator Definition

Reconfigurable hardware comes in with different on chip resources. This means that fixing built-in operators in a language for customizable soft processors will hinder the flexibility offered by reconfigurable hardware. Moreover, different applications require different operations. For example, in FIR filtering, there is vector multiplication (inner product). On the soft-processor, the hardware units for this application can be one of the following: (1) a multiplier and an ALU, (2) a multiply-accumulate unit and an ALU, (3) an inner product. Note that in all of them, there is no operation requirement like subtraction or shifting. Therefore the instruction set of the soft-processor will consist of only one of the following instruction set groups (1) MULTIPLY, ADD, INCREMENT, COMPARE, (2) MULTIPLY_AND_ACCUMULATE, INCREMENT, COMPARE, (3) DOT_PRODUCT.

We propose that the designer should determine how the built-in operators will be. Therefore, while developing the application, the user has the freedom to define the operators. The user picks one of the methods while defining an operator:

- Select an operator from an already-designed-operators library or invoke a module generating function from the related library.
- Define the operator with an HDL that the user selects. RH(+) treats HDL-defined modules as black boxes. Therefore it can handle different types of HDLs simultaneously. It uses interface signals to connect the operator to the data path.
- Write the behavior of the operator with the same language that is used for developing an application. LRH(+) is the language that we propose as a solution. In that case, the basic operations in the operator definition must be introduced to the system by using the methods described above.

The operators can be defined globally as well as locally for each loop or function. This flexibility will give the chance for run-time reconfiguration of the soft-processor, i.e. for each function and loop, the instruction-set and datapath will be reconfigured if it is desired by the user. Each operator must have a name specified by the user. However, the same name can be given to operators of similar nature so that it will make code development much easier. This can be explained with a simple example: Assume that in a function, there is an addition operation acting on scalar variables. Let its name be "+" . Assume that in another function there are two vectors of the same size and we want to realize a vector addition. We can also set its name as "+". It is the job of the compiler to map the same "+" to the related hardware unit. Actually, the front end compiler must attach an attribute to the intermediate format of the user code so that the back end compiler resolves two different "+"s in two different functions. An example is shown in Section 5.

Note that our model leaves the decision about some parameters like the operator's opcode, behavior of the operator, assembly syntax of the operator, register files at which operands are located, wires defining input and output nets, and physical addresses for memory locations to the compiler. Further details about operator definition are explained in Section 4.

Flexible Data Types

In software development, the application developers use variables that have different size. Usually they use keywords like char, int, long to identify the size. However, these keywords might correspond to different bit-size and format in each data type might change in each processor. Moreover for reconfigurable systems, having custom bit-lengths for each variable causes low utilization of the underlying hardware. For example, assume that we declare a counter which has to set a flag whenever it reaches 5. If we use 8-bit Freescale processor and use *unsigned char* data type for each variable, then we have to use either two 8-bit memory locations. Another solution might be usage of only one *unsigned char* variable and partition it so as to access flag and counter in the same variable. This shows that we will use either 8-bit or 16-bit memory locations to manipulate 4 bits (1 bit for flag, 3 bits for maximum value of the counter).

We claim that the user must have the freedom to enter the size of each variable separately, in bits. Moreover, designers dealing with processor customization must consider only the size and format of the variable types, because instruction generation must be carried out by the compiler, which either (1) maps operations of the same type to the same processing unit or (2) maps operations of the same type to different processing units or (3) sets the reconfiguration parameter so that the hardware will be reconfigured for different operation types or operations with different variable size (spatio-temporal mapping). Reconfiguration is a user-specified compiler directive and once it is set, the decision about reconfiguration instances is given by the design evaluation tool. The compiler also decides on the size of the datapath.

Configurability and Self-Retargettable Compiler

Ideally, a reconfigurable processor must have the chance of being reconfigured prior to execution of each instruction. However, reconfiguration is a slower process than computation. If the size of the target architecture allows, compile-time configuration might be sufficient. Otherwise, run-time reconfiguration must be used only when it is necessary. As a result, spatio-temporal scheduling tools must involve in for the optimized area-performance objectives of the design.

Since target architecture is fully configurable, initially there must be no instructions on the processor. The instruction selection tool must identify all the necessary instructions that will be used in the application, i.e. a partitioning must be realized on the design representation graph. Therefore, we claim that current instruction selection tools need to be improved because they only extract the critical instructions. Other nodes on the design representation graph are realized by general purpose instructions of the processor.

After instruction selection, a template for each instruction must be automatically generated so that the compiler will automatically retargets itself for the new instruction set. Hence, the designer will not spend time in retargetting the compiler.

Constraints Setting

An embedded design has to meet several constraints like area, time, energy, etc. Therefore any embedded application design tool must support entry of the global and local constraints. Since instructions are generated specific to the application, behavioral synthesis tools can be used to satisfy both global and local constraints which appear at functions, loops, conditional branches and even basic blocks (i.e. remaining statement blocks).

4. Implementation of RH(+)

The technology used in the implementation of RH(+) IDE is eXtensible Markup Language (XML) due to its capabilities that are noticed in many works in the literature [3, 4]. XML provides us with storing data structurally. Based on XML, we have implemented a new design entry language, LRH(+). Many features like the instruction templates, the operators, the board definition, the graphs, the constraints, the intermediate representation of the application, and the map of the embedded software to the hardware are stored in XML format. Because, outputs of RH(+) IDE is in XML format, it can moved to another application easily via deserialization of them. Besides, thanks to eXtensible Stylesheet Language (XSL), data transformation from XML to the other representations can be achieved via a set of XSL translation rules.

FRH(+)

FRH(+) is the framework implementation of the RH(+) model (Fig. 7.1). It consists of three major levels, namely, the design entry, the design automation and the final design outputs. In the design entry level, the embedded system designer must define the application-specific environment, i.e. the Board Support Package (BSP), and Operator Definitions (OPDEF). They include hardware abstraction required for RH(+) explained in Section 3. Based on these specifications, configuration data is generated automatically. Then, the user can develop the application by using the configuration data and LRH(+) language. The application and the user defined constraints are handled at the design automation level, which meets "efficient mapping of the designed system to the reconfigurable architecture requirement" of RH(+). At this level, the front end compiler generates an enhanced and optimized Control Data Flow Graph (CDFG). The user-defined constraints and reconfigurability parameters

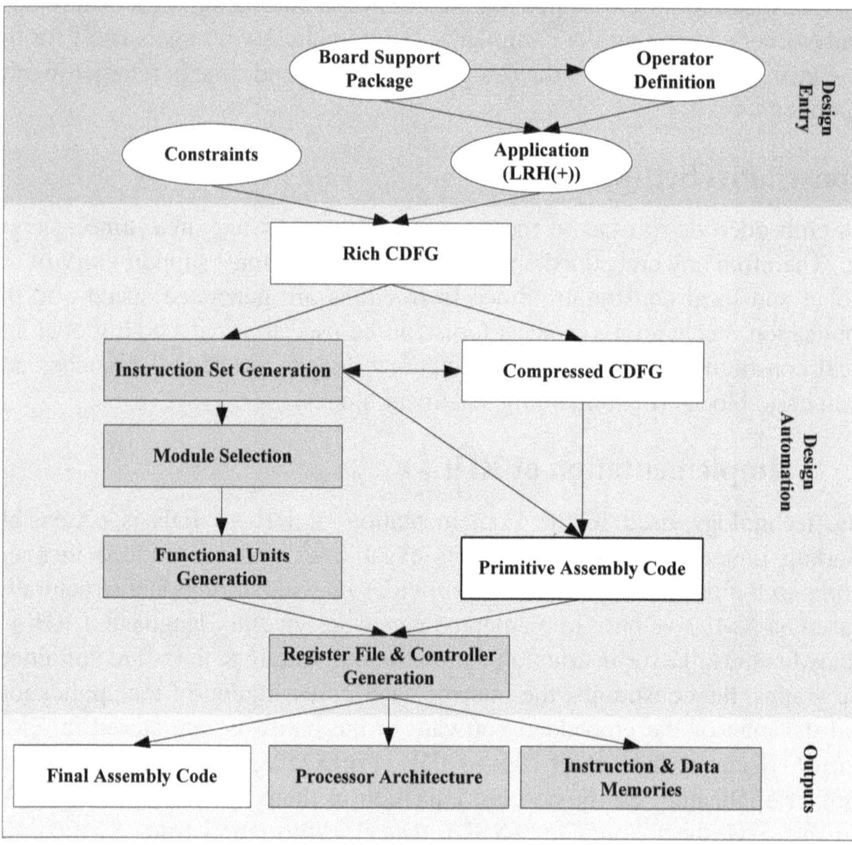

Figure 7.1. FRH(+): framework for the RH(+) model

are used to generate an optimal CDFG. The CDFG is in XML format and used during instruction set generation and module selection. An as-fast-as possible algorithm with area constraints is developed for the instruction selection so as to minimize the overall execution time. The CDFG is compressed with the help of the generated instruction set, i.e. the related nodes are replaced with the selected instructions. Both the compressed CDFG and the instruction set are utilized in producing the primitive assembly code, which assumes infinite resources for the implementation. Instruction set generator also determines the instruction codeword size and prepares the templates for the compiler so that the full assembly code is generated in the final level. As the outputs, we also have the processor architecture, instruction and data memory files. The shaded boxes in Fig. 7.1 indicate the stages which we try to improve further.

LRH(+)

LRH(+) is a language which facilitates sequential and concurrent programming in the same environment [11]. It includes both traditional programming constructs and the constructs special to Custom Hardware (CH) which utilizes reconfigurable and non-reconfigurable parts at the same time on the same device. In addition, it includes structures proposed in FRH(+).

There are two data-types in LRH(+). These are *general* and *array*. The parameters to define a variable with *general* data type are variable name, size in terms of bits, location and initial value of the variable. Location information is retrieved from Templates or definitions at Application Development level of the FRH(+) since embedded system designer can also add location definitions at application level. *Array* data-type requires size of array and initial values of its elements whose sizes are also given in terms of bits. It should be noted that new data-types can be generated by using these data-types. Data types like char, int, long can easily be with *general* type.

In LRH(+), we propose a syntax which is different than traditional languages used in embedded systems programming. LRH(+) compiler also supports traditional syntax indicated as Syntax 1 in Table 7.1. However, we advise the designer to use the new syntax shown as Syntax 2 in Table 7.1, because this syntax has an advantage that each operator can have more than two operands. Since the underlying processor is customizable, it is not unlikely to have a operator with more than two inputs on the data-path.

In LRH(+), each statement in the code has a chance to operate concurrently. Data dependency between the statements dictates the sequential behavior between the statements. However, "?" operator can be used to impose a sequential behavior between two data-independent statements. For example, consider program segment in Fig. 7.2a. Assuming that each statement has identical execution times, the execution order of the segment is found as it is shown

Table 7.1. Syntax of statements and conditions

Tag	Syntax 1	Syntax 2
Statement	c = a + b;	.=(c, .+(a, b));
	d = func1(c);	.=(d, .func1(c));
	e = d/c*4;	.=(e, .*(./(d, c), 4));
Condition	(a > b) && (c < d)	.&&(.>(a,b), .<(c,d))

Program segment	Execution order
*l*1: .=(c,.+(a,b));	*s*1: *l*1, *l*4
*l*2: .=(d,.*(c,e));	*s*2: *l*2, *l*5
*l*3: .=(f,.+(d,k));	*s*3: *l*3
*l*4: .=(g,.−(a,b));	
*l*5: .=(h,.+(g,i));	

(**a**)

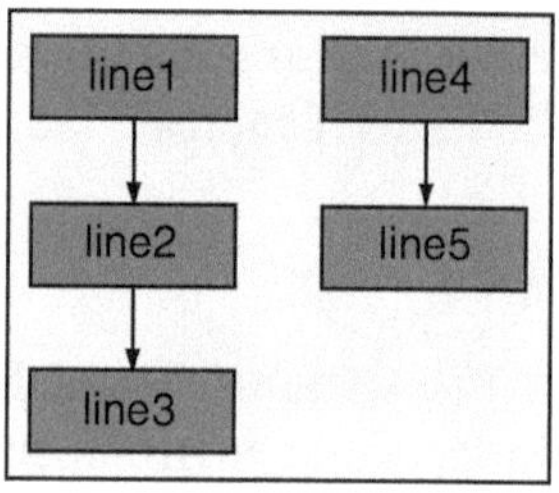

(**b**)

Program segment	Execution order
*l*1: .=(c,.+(a,b));	*s*1: *l*1
*l*2: .=(d,.*(c,e));	*s*2: *l*2
*l*3: .=(f,.+(d,k));	*s*3: *l*3
*l*4: ?.=(g,.−(a,b));	*s*4: *l*4
*l*5: .=(h,.+(g,i));	*s*5: *l*5

(**c**)

Figure 7.2. Concurrent and sequential flow declaration in LRH(+): (**a**) A concurrent program segment. (**b**) Data-flow graph. (**c**) Introducing sequentiality without altering data-flow

in Fig. 7.2a. In this figure, *s* letter represents execution order. However, if a sequential behavior is desired by the user then "?" operator must be perpended to line *l*4, shown as modified code in Fig. 7.2b. As a result, execution becomes sequential.

Constraints and Configurability

In our implementation, global and function level area and performance constraints can be handled. Reconfigurability is a global option. Once it is set, the compiler decides on which module(s) will be reconfigured by taking constraints into consideration.

Instruction Selection and Generation

Instruction selection is done by a fuzzy expert system based on FuzzyClips [16]. By using the CDFG, it extracts recurring subgraphs and sets of subgraphs that can execute concurrently. Then using behavioral synthesis as a subroutine, it evaluates the area and time costs of operations, subgraphs and concurrent candidates. This information is passed to the fuzzy expert system and the set of selected instructions are passed to the graph evaluator. The best set is selected based on the constraints and the adaptive threshold. The selected instructions are marked on the CDFG and the procedure repeats itself from the extraction of subgraphs phase until all nodes are marked.

After the instructions are selected, the CDFG is modified so that each recurrent subgraph or each set of concurrent subgraphs selected as an instruction is compressed to a node. The related hardware modules are generated and a node-based decision is taken for reconfiguration by using data dependencies and constraints.

A register file is generated based on the assumption that each instruction fetches the operands from a register file and writes the result to the register file. The basic reason for this assumption is that RAM blocks in FPGA support single read-write in each access. Register file can contain registers with different widths.

An instruction template is generated for each instruction. The template generator assumes that the operands have to be moved from the memory to the registers and the output can reside in register(s). For example, consider the statement .=(t4,.+(t4,t3)) in Fig. 7.5. It is simply the addition of two numbers (t4 and t3) and the result is written back in t4. Assume that t4 and t3 are defined as 8 bit scalars using General data-type. Assume that a scalar adder and store instruction is selected as an instruction for this operation. Then the related template is generated as shown in Fig. 7.3.

MOV is a keyword that realizes the movement from memory to the register file. There are as many MOVs as the number of inputs. Register names starts with either RG or RA. RG registers stand for general register type. These are in the block RAMs of the FPGA. RA registers represent array type registers.

```
% scalar addition .+(t4,t3)
MOV RG8_0 op1
MOV RG8_1 op2
PLUSG8G8G8_0 RG8_0 RG8_0 RG8_1
% scalar store .=(t4,...)
MOV op2 RG8_0
```

Figure 7.3. A sample template generated by the compiler for add

These registers are distributed in the FPGA. A bulk move from distributed memory to distributed register file is realized when RA is the target register. The number following RG or RA represents the word-length of the register. The numbers separated by "_" represent the instance of the register. For example, there are two RG8 registers in the sample template, namely RG8_0 and RG8_1. The operands of the instruction starts with op1, op2,... and they are as many as inputs of the instruction. Each instruction starts with its node name. It should be noted that the compressed nodes are given special names by the instruction generation module. If an operation does not contain alphabetical characters, special names are given in a way not to coincide with other instruction names. For example, PLUS is given for "+" operand. It should be noted that the compressed nodes are given special names by the instruction generation module. Type and size of the registers are appended to the node name prevent misinterpretation of the instruction.

Note that the register and operand instances, and MOV instructions are dummy, i.e. they are replaced with the correct instances by the compiler. Actually, the instances of all registers and operands start with 0 in a template of an instruction.

Based on the number of instructions and number of inputs, the machine code of instructions are automatically generated with a predefined format. In addition to the instructions covered in the CDFG, instructions like MOV, RECONFIGURE, NOP, STOP are always included to the set.

Compiler

The compiler firstly converts the modified CDFG to a parse tree. Then it calculates the minimum size of the register file by using the templates. This size can be increased by the compiler if the register size does not suffice during code generation. Note that reduction is always possible by tightening area constraints.

During the traversal of the parse tree two lists for registers are updated to monitor status of the registers. The first list keeps track of busy registers. A busy register means a register which is occupied at that time. The second list keeps track of the operand which makes the register busy. The compiler generates the assembly code for each node in the parse tree if its siblings have already been processed. Therefore the output of each sibling is searched in the second list so as to see whether it appears. If it is the case, then the related MOV instruction in the template is skipped and the busy flag in the first list is set. Otherwise, an empty register is searched in the first list so as to replace the related dummy register with the empty register. Then the status of the selected register is changed from empty to busy and the second list is updated with the output of the sibling. As soon as the related code is generated, the status of the registers using the outputs of the siblings are reset to free. There exists no case where all registers are busy because the size of register file is calculated prior to compilation.

After the assembly code is generated, then machine code is generated by using the instruction machine code explained in Section 4.

5. Illustrative Example

In this section we consider a motion intensity calculator and implement it using RH(+)IDE. In BSP, we define two RAMs to store the general and array variables separately, and a serial bus to put the MPEG-7 representative of each block motion block to the array memory. For this application, we defined the operators shown in Table 7.2 in OPDEF. Six general-type (average, intensity, index, t3, t4, 12) and five array-type (motionVectorX, motionVectorY, T1, T2, T) variables are defined.

Figure 7.4 demonstrates the code we write in RH(+). This function gets two arrays, *motionVectorX* and *motionVectorY* with length of 12 and then calculates the square of each item in these arrays using "square" operation and the results are written to *T1* and *T2* arrays, respectively. Next, in a loop, each item in *T1* and *T2* are added, square root of the sum is calculated and the result is added to *t3*. After the loop, *average* is obtained by dividing *t3* to length of arrays, i.e. 12. At the same expression the two square arrays are added and saved in array *T*. Then the root of each value in *T* is assigned to *t3* and then inside an IF statement, it is checked whether the amount of *t3* is bigger than or equal to *average*, by using ">=" operator. If it is bigger then *t3* is added to *t4*. Afterwards, the intensity variable is calculated as the ratio of *t4* to *12* (length of arrays).

Table 7.2. Operation defined for motion intensity calculator example

OP Name	Input1 type	Input2 type	Output type	Description
>=	General	General	General	Bigger or equal
<	General	General	General	Smaller
+	General	General	General	Integer addition
*	General	General	General	Integer multiplication
/	General	General	General	Integer division
Root	General	–	General	Root
++	General	–	General	Increment
Square	Array	–	Array	Square amount of each index in array
+	Array	Array	Array	Addition of each index in two arrays
[ ]	Array	General	General	Amount of the special index in a array

```
Function;MotionIntensityCalculator:F1;F1;
Expressions;exp1;exp1;
    .=(T1,.square(motionVectorX));
    .=(T2,.square(motionVectorY));
ExpressionsEnd
Loop;L1;L1;
    .<(index,12);
    Expressions;exp2;exp2;
        .=(t3,.+(t3,.root(.+(.[](T1,index),.[](T2,index)))));
        .=(index,.++(index));
    ExpressionsEnd
LoopEnd
Expressions;exp3;exp3;
    .=(average,./(t3,12));
    .=(T,.+(T1,T2));
    .=(index,0);
ExpressionsEnd
Loop;L2;L2;
    .<(index,12);
    Expressions;exp4;exp4;
        .=(t3,.root(.[](T,index)));
        .=(index,.++(index));
    ExpressionsEnd
    If;IF1;IF1;
        .>=(t3,average);
        Expressions;exp5;exp5;
            .=(t4,.+(t4,t3));
        ExpressionsEnd
    IfEnd
LoopEnd
Expressions;exp6;exp6;
    .=(intensity,./(t4,12));
ExpressionsEnd
FunctionEnd
```

Figure 7.4. LRH(+) code of motion intensity calculator

The produced .cdfg file of this design is about 2000 lines and it is inappropriate to demonstrate this file in this paper. However, we try to show a part of this file in Fig. 7.5. This figure indicates the generated code for expressions *".+(t4,t3);"* and *".+(T1,T2);"* on lines 27 and 15, respectively in Fig. 7.4. The compiler also generates an assembly file which consists of 69 lines of code, as shown in Fig. 7.6. In the assembly code, we simplified the instruction mnemonics for readability. Figure 7.7 shows the hardware implementation of SQRA instruction which takes the square of all entries in an array stored in the distributed memory of the FPGA. This instruction is generated in second and fifth lines of the assembly code in Fig. 7.6. In the unit, we consider that a scalar square-taking unit is utilized for each entry in the array. Hence there

```
-<component.type>
  <!–specifications–>
  <Type>General</Type>
  - <GeneralTypeData>
    <Length>8</Length>
    <Value>0</Value>
    <Location>MEM1</Location>
    <refName>t4</refName>
  </GeneralTypeData>
- </component.type>
************************
- <component.type>
  <!–specifications–>
  <Type>General</Type>
  - <GeneralTypeData>
    <Length>8</Length>
    <Value>0</Value>
    <Location>MEM1</Location>
    <refName>t3</refName>
  </GeneralTypeData>
- </component.type>
************************
- <SubExpression>
  <Number>754</Number>
  <Name>SE85</Name>
  <!–specifications –>
  <OpName>+</OpName>
  - <OperandType>
    <Location>MEM1</Location>
    <Type>General</Type>
  </OperandType>
  - <OperandNames>
    <string>+</string>
    <string>t4</string>
    <string>t3</string>
  </OperandNames>
</SubExpression>
```

```
-<component.type>
  <!–specifications–>
  <Type>Array</Type>
  - <GeneralTypeData>
    <Length>8</Length>
    <Value>0</Value>
    <Location>MEM1</Location>
    <refName>T1</refName>
  </GeneralTypeData>
- </component.type>
************************
- <component.type>
  <!–specifications–>
  <Type>Array</Type>
  - <GeneralTypeData>
    <Length>8</Length>
    <Value>0</Value>
    <Location>MEM1</Location>
    <refName>T2</refName>
  </GeneralTypeData>
- </component.type>
************************
- <SubExpression>
  <Number>745</Number>
  <Name>SE660</Name>
  <!–specifications –>
  <OpName>+</OpName>
  - <OperandType>
    <Location>MEM1</Location>
    <Type>Array</Type>
  </OperandType>
  - <OperandNames>
    <string>+</string>
    <string>T1</string>
    <string>T2</string>
  </OperandNames>
</SubExpression>
```

Figure 7.5. Intermediate CDFG code generated for scalar and array adders

are 12 units. Since the arrays are located in distributed memory of the FPGA, the entries of the selected array can be accessed simultaneously, in the same clock cycle. The array selection is decided by the array pointer stored in R0. By using a multiplexer we can select the array that we want to calculate the square. The result of the calculation is sent to the same array, i.e. source and target arrays are the same as indicated in Fig. 7.6. The demultiplexer is used for selecting the target array. This unit is also combined with ADDA instruction. However, only the parts needed for square unit is cited in the figure to be illustrative enough and to prevent complexity.

```
MOVA R0 motionVectorX
SQRA (R0) (R0)
MOV T1 R0
MOVA R0 motionVectorY
SQRA (R0) (R0)
MOV T2 R0
L0: MOV R0 index
MOV R1 12
CMP R0 R1
JGE L2
MOV R0 1
Jump L3
L2: MOV R0 0
L3: CMP R0 0
JE L1
ARRINDEX R0 T1 index
ARRINDEX R1 T2 index
ADD R0 R0 R1
root R0 R0
MOV R1 t3
ADD R1 R1 R0
MOV t3 R1
MOV R0 index
ADD R0 R0 1
MOV index R0
Jump L0
L1: MOV R0 t3
MOV R1 12
DIV R0 R0 R1
MOV average R0
MOVA R0 T1
MOVA R1 T2
ADDA (R0) (R0) (R1)
MOV T R0
MOV R0 0
MOV index R0
L4: MOV R0 index
MOV R1 12
CMP R0 R1
JGE L6
MOV R0 1
Jump L7
L6: MOV R0 0
L7: CMP R0 0
JE L5
ARRINDEX R0 T index
root R0 R0
MOV t3 R0
MOV R0 index
ADD R0 R0 1
MOV index R0
MOV R0 t3
MOV R1 average
CMP R1 R0
JG L10
MOV R0 1
Jump L11
L10: MOV R0 0
L11: CMP R0 0
JE L8
MOV R0 t4
MOV R1 t3
ADD R0 R0 R1
MOV t4 R0
L8: Jump L4
L5: MOV R0 t4
MOV R1 12
DIV R0 R0 R1
MOV intensity R0
```

Figure 7.6. Assembly code generated for MotionIntensity function

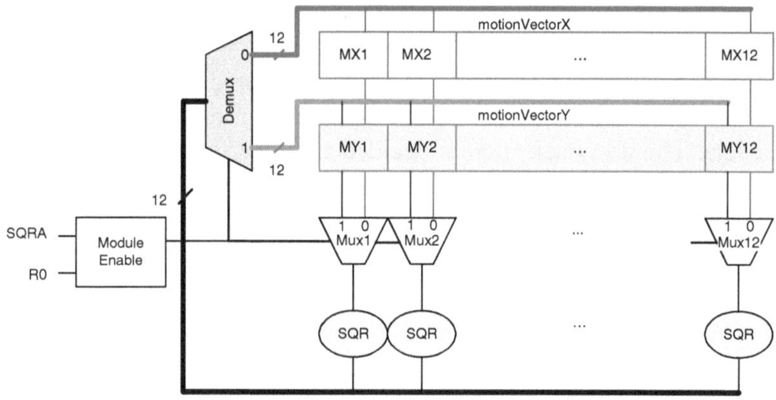

Figure 7.7. The hardware schematic for a square unit of array type

6. Conclusions and Future Work

In this paper, we have proposed a model, RH(+), for embedded system development on run-time reconfigurable hardware. Based on this model, we have implemented a design environment which includes a framework FRH(+), a new design language LRH(+), an instruction selection and generation tool and a compiler. The LRH(+) language enables the user to develop the application with flexible data types and operators. It generates an extended CDFG which is an intermediate representation in XML format. The instruction selection/-generation toolbox generates the the complete instruction set from the CDFG. The compiler automatically retargets itself for the new instruction set. As the current work we try to implement a runtime reconfiguration of the processor, i.e. to activate RECONFIGURE instruction. Another issue is the automatic generation of the whole processor with its control unit and data path being connected to. Finally, we are dealing with the integration of these tools in the same development environment.

Acknowledgment

This work has been partially supported by the Turkish Foundation of Science and Technology, under Kariyer Research Project Program (pr.no:104E038). We also would like to thank to anonymous reviewers who have helped us to improve the quality of this article.

References

[1] Altera Corporation, Nios embedded processor, 2007.

[2] J.A. Bower, et al., A java-based system for fpga programming. *FPGA World Conference*, 2008.

[3] W.O. Cesario, et al., Colif: a design representation for application-specific multiprocessor socs. IEEE Des. Test **18**, 8–20, 2001.

[4] F.P. Coyle, M.A. Thornton, From UML to HDL: a model driven architectural approach to hardware-software co-design. Information Systems: New Generations Conference (ISNG), vol. 1., 88–93, 2005.

[5] R. Damasevicius, A subset-based comparison of main design languages. Inf. Tech. Cont. Technol. **1**, 49–56, 2004.

[6] A. Fauth, J, Praet, M. Freericks, Describing instruction set processors using nML. *EDTC'95*, pp. 503–507, 1995.

[7] Freescale Semiconductor Inc., CodeWarrior development tools, 2007.

[8] R.E. Gonzalez, Xtensa: a configurable and extensible processor. IEEE Micro **20**, 60–70, 2000.

[9] A. Halambi, et al., A customizable compiler framework for embedded systems. *SCOPES*, 2001.

[10] T. Kambe, A. Yamada, M. Yamaguchi, Trend of system level design and an approach to c-based design. Microelec. J. **33**, 875–880, 2002.

[11] B. Kurumahmut, RH(+): the model for high-level embedded system design on run-time reconfigurable hardware. Master's thesis, Bogazici University, Computer Engineering, 2007.

[12] B. Mei, et al., Architecture exploration for a reconfigurable architecture template. IEEE Des. Test **22**, 90–101, 2005.

[13] S.J. Mellor, Executable and translatable UML, 2003.

[14] S.V. Nikolaos, M. Konstantinos, *System Level Design of Reconfigurable Systems-on-Chip*. Springer, Dordrecht, The Netherlands, 2005.

[15] Open SystemC Initiative, www.systemc.org., 2009.

[16] R.A. Orchard, Fuzzyclips version 6.04a Users' guide, 1998.

[17] I.R. Quadri, S. Meftali, J.L. Dekeyser, High level modeling of dynamic reconfigurable fpgas. International Journal of Reconfigurable Computing, vol. 2009, Article ID 408605, 2009. doi:10.1155/2009/408605.

[18] C. Raistrick, P. Francis, J. Wright, *Model Driven Architecture with Executable UML(TM)*. New York, NY, Cambridge University Press, 2004.

[19] O. Schliebusch, et al., Architecture implementation using the machine description language lisa. ASP-DAC'02, pp. 239–244, 2002.

[20] Silicon Hive, HiveCC software development kit v3.5., 2009.

[21] SpecC Technology Open Consortium, SpecC language reference manual, version 2.0., 2002.

[22] Synfora Inc., www.synfora.com., 2009.

[23] Target Compiler Tech., CHESS/CHECKERS: a retargetable tool-suite for embedded processors. Technical report, Belgium, 2003.

[24] S. Vassiliadis, et al., The MOLEN polymorphic processor. IEEE Trans. Comp. **53**, 1363–1375 (2004).

[25] Xilinx Inc., www.xilinx.com., 2007.

[26] P. Yiannacouras, J. Rose, J.G. Steffan, The microarchitecture of fpga-based soft processors. CASES'05, pp. 202–212, 2005.

[27] V. Zivojnovic, S. Pees, H. Meyr, LISA-machine description language and generic machine model for HW/SW co-design. *IEEE Workshop on VLSI Signal Processing IX*, pp. 127–136, 1996.

Chapter 8

A SYSTEMC SUPERSET FOR HIGH-LEVEL SYNTHESIS

Maxim Smirnov and Andres Takach

Mentor Graphics Corporation, 8005 SW Boeckman Rd., Wilsonville, OR 97070

Maxim_Smirnov@mentor.com

Abstract Today's System-on-a-Chip (SoC) designs are becoming more and more complex. This creates demand for system level Computer-Aided-Design (CAD) tools that allow early hardware and software co-development, hardware and software performance evaluation and fast system-level modeling. The latest achievements in this area are mostly due to wide SystemC adoption and the recent introduction of the TLM 2.0 standard. The complexity of today's SoC designs makes High-Level Synthesis (HLS) an important part of modern design flows.In this paper, we analyze two basic approaches to HLS user input that exist today – sequential ANSI C/C++ and SystemC Synthesizable Subset. Based upon the results of our analysis, we propose a new concept – SystemC Synthesizable Superset – a solution that combines advantages of both approaches (ANSI C/C++ and SystemC), but does not inherit their disadvantages. Our HLS-oriented concept features a predefined library of HLS modules, user threads with implicit timing specification and a set of TLM 2.0 compatible interfaces. In addition, our HLS objects allow for various levels of simulation abstraction (or timing accuracy), such as cycle-accurate at transaction boundaries (CATB), approximately-timed and loosely-timed modeling. Simulation abstraction levels can be switched without the need to rewrite the user system specification that determines high flexibility level of our solution. In conclusion, we demonstrate a simple synthesizable video system and compare the simulation speeds at different abstraction levels.

Keywords: HLS, SystemC, TLM 2.0, interfaces, scheduling, synthesis constraints, time annotation points, CATB, approximately timed, loosely timed

1. Introduction

Electronic system level (ESL) based design is gaining adoption in industry mainly in two areas: hardware design from high level specifications, and system-level validation and analysis using transaction-level modeling (TLM)

D. Borrione (ed.), *Advances in Design Methods from Modeling Languages for Embedded Systems and SoC's*, Lecture Notes in Electrical Engineering 63,
DOI 10.1007/978-90-481-9304-2_8, © Springer Science+Business Media B.V. 2010

techniques. The convergence of these ESL design methodologies for hardware and embedded software is reshaping the way systems are designed. The adoption of hardware design based on high-level synthesis provides an implementation path to generate optimized RTL from abstract un-timed or partially timed models. The release of the SystemC TLM 2.0 standard ([9]) provides new tools for modeling components of the systems at various levels of timing accuracy and simulation speed. For example, loosely-timed (LT) models provide fast models of hardware that enable early development of firmware and embedded software. Approximately-timed (AT) models provide more accurate timing at the expense of simulation speed.

The key benefit of using HLS comes from its high-level of abstraction with a clear separation between algorithm and implementation constraints that define cycle timing, hardware interfaces and overall micro-architecture of the generated hardware. The functionality of a block or subsystem can typically be specified at a high level of abstraction where the cycle timing is not an integral part of the intrinsic functionality (algorithm). The cycle timing is added by synthesis based on implementation constraints that include interface synthesis. When specifying an entire system, however, the designer is providing certain macro-architectural details of its parts considering that cycle timing of individual blocks may affect the functionality of the overall system. The ultimate goal is to be able to specify, verify and synthesize the system in its entirety. The requirement to achieve that goal is to provide an executable specification of the system that can be validated, synthesized, and the resulting hardware verified against the executable specification. An additional requirement on the executable specification is that it needs to be easily configurable for different levels of timing accuracy (for example LT) to enable using it for firmware or embedded software development ([3]).

Behavioral specification styles for HLS based on sequential ANSI C/C++ deliver a high level of model abstraction and portability, but do not allow parallelism necessary for describing design hierarchy of concurrent systems, arbitration and time-dependent algorithms, which mostly limits applicability of such styles to block or subsystem-level HLS.

Behavioral specification styles based on SystemC ([6, 8, 12]) that are targeted for HLS make extensive use of wait-on-clock statements. These specification styles do not provide a sufficient separation of function from cycle timing as wait statements constrain all I/O assignments that precede the wait from all the I/O assignments that succeed the wait. The use of waits unnecessarily constrains I/O and inhibits HLS from being able to perform important optimizations such as loop pipelining and interface optimizations. Unless the design is latency insensitive [2], the simulation of the entire SystemC specification and the generated RTL are not guaranteed to match, even though all the blocks of a system are synthesized correctly. Finally, the use of waits implies context switching in simulation, causing significant increases in runtime. All of these issues make the use of such behavioral models impractical for system-level design.

This paper addresses those challenges and proposes a novel modeling methodology that is TLM 2.0 compliant and provides improved "speed vs. accuracy" trade-offs while preserving a high level of abstraction for high-level synthesis.

2. SystemC Synthesizable Superset

Our SystemC Synthesizable Superset overcomes the difficulties typical to sequential C/C++ and SystemC HLS specifications by defining a special high-level synthesis layer on top of the SystemC and TLM 2.0 libraries. This layer allows the behavioral portion of a specification (user source code) to remain un-timed and unambiguously defines target synthesis hardware architecture and timing. Similar approaches can be found in [4, 5, 10], which all, in some form, establish a connection between an input SystemC model and the target hardware – the output of high-level synthesis. This article attempts to further refine and develop this idea.

The Synthesizable Superset is a C++ library of classes. All the components in this library are built with the assumption that a user specification is subject to later high-level synthesis employing static scheduling, and the result of this synthesis step is hardware with known structure and certain timing properties. This, in our opinion, should be seen as a major advantage of our solution over traditional "raw" SystemC-for-HLS design specification, where the only available instrument to describe timing is wait statements. As shown later in this article, wait statements do not provide sufficient means to simulate the effects of loop pipelining and unrolling, and furthermore may allow a user to describe a design with timing characteristics unreachable in principle (illegal schedule).

The key components of the SystemC Synthesizable Superset are:

- Modules
- Interfaces
- Threads and schedule objects associated with them
- Shared Variables

HLS Modules

HLS modules are design hierarchy leaf elements – blocks or sub-blocks of SoC designs. To describe modules, our SystemC Superset provides a base class called *hls_module*, which should be used by all user-defined modules as their parent class. The hls_module base class, in its turn, is derived from the standard SystemC *sc_module* and, therefore, nested SystemC hierarchy and binding

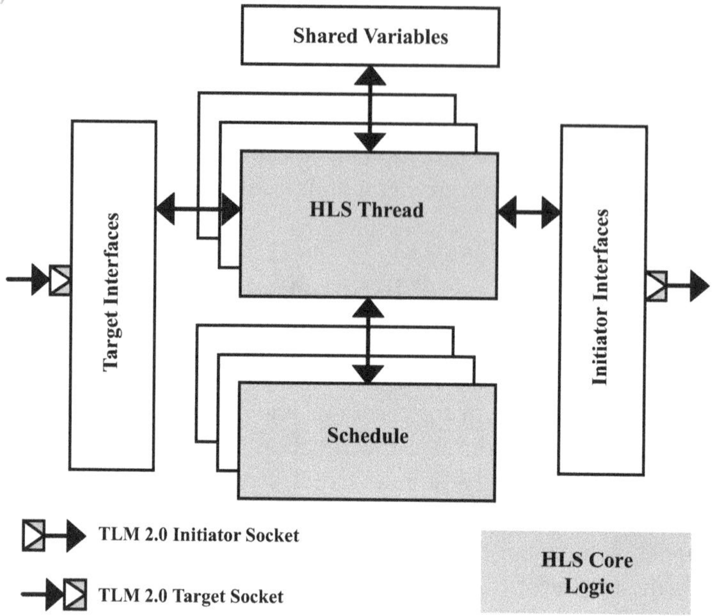

Figure 8.1. HLS module structure

mechanism are fully supported. The behavioral portion of a user specification for synthesis should be defined in threads of derived classes. Modules with multiple threads are allowed. User threads communicate with other threads that belong to different modules through HLS Interfaces. Threads that belong to the same module can communicate with each other by means of HLS Shared Variables. All the threads of an HLS module are synchronized to one, and only one, clock source, though interface components allow re-synchronization logic between different clock domains. The structure of an HLS module is presented in Fig. 8.1.

HLS Interfaces

HLS Interfaces are custom objects with pre-built RTL implementation and corresponding SystemC simulation models. Synthesis of interface components from a high-level description into a low-level signal protocol (so-called "transactor synthesis"), is not precluded by the proposed solution; however it is outside the scope of this article.

For inter-module connection, HLS interfaces provide TLM 2.0 initiator and target sockets or wires, i.e. interfaces based on sc_signal. User threads communicate with interface objects by calling read and write interface access methods, and both methods can be either blocking (in SystemC definition) or non-blocking. Interface access methods correspond to hardware ports in a synthesized design, through which the synthesized thread core logic is connected to the pre-built module's interface logic. Please note a major difference with "raw" SystemC – core logic (user code) can communicate with an interface component only in a limited number of predefined ways that are supported by a synthesis tool, thus helping to avoid "non-synthesizable" constructs in a user's code.

Interface Object Types. There are three basic types of HLS interface objects:

- Wire interface objects
- Initiator interface objects
- Target interface objects

Initiator and Target interfaces are distinguished by the TLM 2.0 socket type they use. In addition to read and write access methods, Target Interfaces can provide conditions (such as "interface has more then one transaction to process") and interface events, which can be used by Target threads as described further in the article. Target interface events must not to be confused with SystemC events. Events and conditions in our approach assume non-ambiguous implementation after high-level synthesis. From the hardware perspective, conditions represent Boolean inputs of thread core logic that can be sampled in a certain clock cycle of the schedule. Events are synthesized as pulses that have width of a single clock cycle.

Initiator and Target Interfaces are shown in Fig. 8.2.

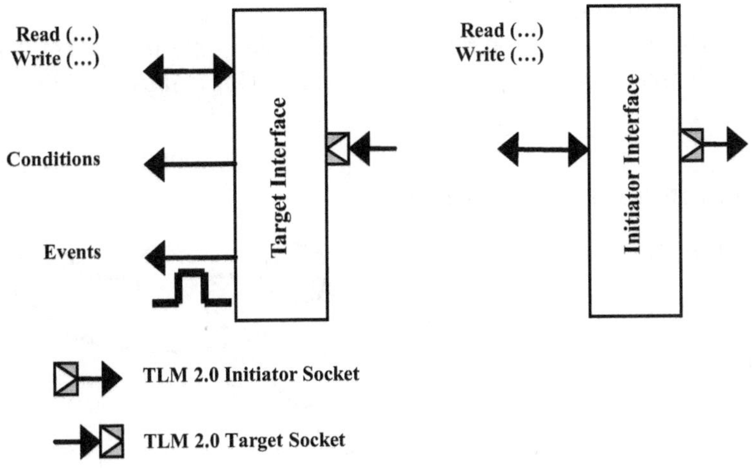

Figure 8.2. Initiator and target interface objects

From a complexity point of view, interfaces can be classified into Primitive Interfaces, which are typically supplied with an HLS tool, and Custom Interfaces, which represent more sophisticated hardware and, as a rule, are supplied by a tool user. A list of Primitive Interfaces may include:

- Wire Interfaces
- Simple Interfaces with and without handshaking
- Message Interfaces – interfaces that allow modules to exchange messages typically a pulse of one clock cycle width accompanied by a message

Examples of Custom Interface Components are:

- Interfaces with internal FIFOs/queues
- Serial-to-parallel and parallel-to-serial converters
- System Busses

HLS Threads

The SystemC Superset supports two types of user threads: Initiator Thread and Target Thread.

An Initiator (or free-running) Thread represents an HLS datapath, which is controlled by a datapath finite control state machine (DPFSM) that has only active states. In other words, a hardware block synthesized from an Initiator Thread becomes active immediately after the reset signal is de-asserted and continues to run through its full execution cycle until its DPFSM reaches a state that requires blocking read or write. This is illustrated in Fig. 8.3. A similar approach can be found in [7].

Target Threads, on the contrary, can be activated only if an event happens or a condition becomes true on a target interface. This is illustrated in Fig. 8.4,

```
void MyInitiatorThread ()
{
  for (;;)
  {
    /* ... */
  }
}
```

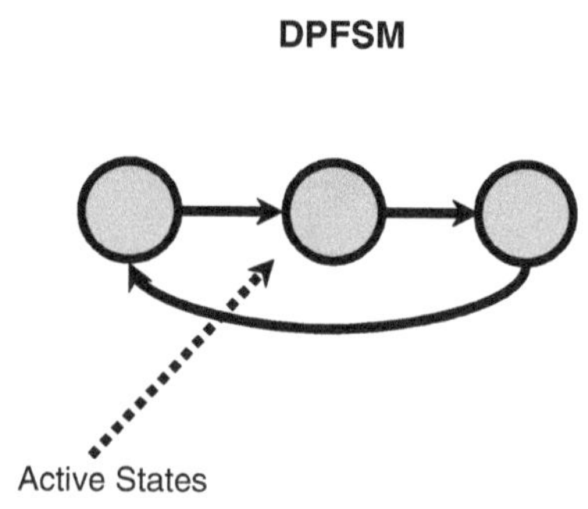

Figure 8.3. Initiator thread

```
void MyTargetThread ()
{
 for (;;)
 {
  wait (intf1.get_event());
  do
  {
     /*...*/
  } while (intf1.get_cond())
 }
}
```

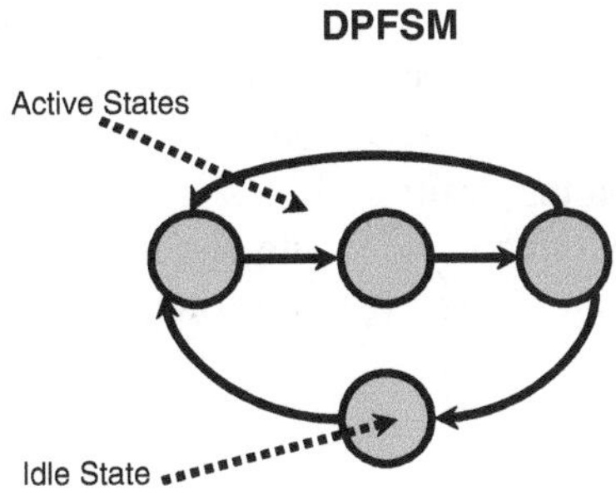

Figure 8.4. Target thread

where the thread is activated only when a condition on the interface intf1 becomes true and remains active until the condition becomes false. A thread activated by an event can be described in a similar way with only one difference – it does not need a while loop, only a wait statement. Events and conditions can be combined in "AND" and "OR" type thread sensitivity lists, similar to sensitivity lists of combinational processes in the synthesizable VHDL and Verilog subsets.

HLS Shared Variables

Shared Variables provide a fast way for two threads of the same module to communicate with each other without having to go through interface components. Typical usage scenarios of shared variables include cases in which one thread only writes into a shared variable while the other thread only reads it; or one thread increments the variable while the other thread decrements it, and so forth.

All shared variables in a user specification (data members of *hls_module*) should be derived from the *hls_shared_variable* base class, which defines a pure virtual *resolve()* function that *must be* defined in the derived class. This function resolves conflicts (race conditions) that may occur when two different threads are trying to write a shared variable in the same clock cycle. This *resolve()* function is used by an HLS tool to build necessary set-reset logic.

Schedule Object and Timing Annotation Points

As we previously pointed out, user threads do not permit usage of *wait* statements. Instead, thread timing is defined by using timing annotation points and assembling them into a thread's execution schedule. Timing Annotation Points (TAPs) include provisions for pipelining and loop unrolling – a feature that is unavailable with *wait* statements.

A schedule object associated with each user thread assures that thread's timing characteristics correspond to a *legal static schedule* that can be potentially produced by HLS tools. Note that this schedule is not necessarily achievable given the desirable clock frequency. Usage of TAPs is illustrated in Fig. 8.5.

A thread's schedule should be assembled out of TAPs at a module's construction time. This is shown in Fig. 8.6.

```
void MyThread ()
{
  // Main Loop
  main_loop_start.execute ();
  for (;;)
  {

    /* Code before the inner loop*/

    // Inner Loop
    inner_loop_start.execute ();
    for (int v=0; v<IMAGE_V_SIZE; v++)
    {
      A = Intf1.read ();
      Intf2.write (ConstA*A);
      inner_loop_iteration.execute ();
    }

    /* Code after the inner loop*/

    main_loop_iteration.execute ();
  }
}
```

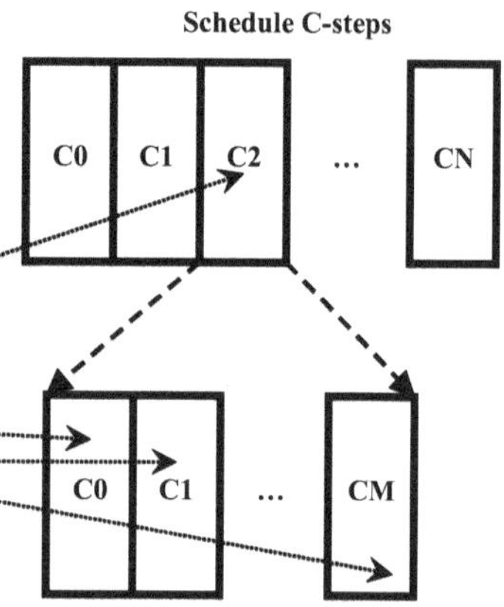

Figure 8.5. Adding timing annotation points into the user code

```
// the following code should be
// present in the module constructor

// the schedule initially includes
// main_loop_start time annotation point

schedule.add_tap (main_loop_start, IIm, Lm);
schedule.add_tap (inner_loop_start, IIi, Li 2, false);
schedule.add_tap (Intf1, 0);
schedule.add_tap (Intf2);
schedule.add_tap (inner_loop_iteration, 4);
schedule.add_tap (main_loop_iteration, 8);

schedule.validate ();
```

The inner loop is scheduled in C-step 2 of the main loop

This offset does not create a synthesis constraint

I/O read should be scheduled in C-step 0 of the inner loop

The default position for the I/O write is at the end of the corresponding loop

Figure 8.6. Constructing the schedule

3. Timing Accuracy Levels

Modules designed with SystemC Synthesizable Superset can be simulated at various timing accuracy levels. A thread's schedules and interface models assure that the appropriate simulation accuracy level is used. Switching between the accuracy levels does not require any user code modification – the switch between the accuracy levels is fully handled by TAPs, schedule objects and interfaces. The current implementation extends the list of simulation timing accuracy levels defined in [9].

Approximately-Timed Simulation Mode

In this mode, thread schedules and interfaces use *wait (N*ClockPeriod)* – waiting for SystemC simulation time – to correctly model timing of thread schedules. Interfaces employ TLM 2.0 non-blocking transport for simulation of interface protocols. Our implementation of this simulation mode uses TLM 2.0 Payload Event Queues.

Cycle-Accurate at the Transactions Boundaries (CATB) Simulation Mode

This simulation mode, although not defined in TLM 2.0, also uses TLM 2.0 non-blocking transport mechanism. The main difference with Approximately-Timed modeling is that at the CATB level all TLM 2.0 non-blocking transport calls are synchronized to clock, the threads are sensitive to a clock edge and thread schedules and interfaces exercise *wait (N)* calls – waits for a given number of clock cycles. This abstraction level is an interim level between cycle-accurate and approximately-timed modeling. Its main purpose is hardware verification after HLS. More information on CATB modeling and verification can be found in [1, 11].

Loosely-Timed Simulation Mode

Our implementation of Loosely-Timed Simulation Mode uses global time quantum and local quantum keepers in every thread's schedule as defined in [9]. In order to allow threads to run independently of each other with their local clock running ahead of the SystemC clock, we use virtual buffers in the interface components that store interface transactions along with their corresponding time stamps.

In order to advance their local clocks, thread schedules and interfaces exercise *m_qk.inc(N*ClockPeriod)* calls – local quantum keeper increment – instead of wait (), thus minimizing the frequency of context switching between the threads, and accelerating simulation.

Context switches may occur only if one of the following occurs:

- Virtual buffer overflows or underflows.
- Thread explicitly calls for local clock synchronization.

In Loosely-Timed Simulation Mode, interfaces employ TLM 2.0 blocking transport calls with time annotation to transfer data between modules.

Un-timed Simulation Mode

The SystemC superset can potentially support fully un-timed simulation for a certain class of designs, although this accuracy level is not a part of the TLM 2.0 standard.

4. Design Flow

Pre-Synthesis Models are available for a designer before High-Level Synthesis takes place. In this model, thread schedules reflect only synthesis constraints and early schedule estimations. The purpose of Pre-Synthesis Models is source code debugging and validation of the synthesis constraints. If required, some schedule parameters may be selected randomly in order to help prove design schedule-independent behaviour.

Post-synthesis models contain actual thread schedules and are meant for final verification and schedule debugging.

The design flow is shown in Fig. 8.7. Please note that thread source code versions before and after HLS are not necessarily equivalent – the latter one should account for all optimizations that took place during synthesis, such as I/O reordering and loop merging.

5. Results

In order to demonstrate usage and scalability of our models for High Level Synthesis and analyze "timing accuracy vs. performance" trade-offs these models deliver; we used a small video design presented in Fig. 8.8. The system included three modules: Video Data Generator, Downscaler and System Memory Model. All modules are single threaded.

The Video Data Generator is an initiator thread that produces random 8-bpp black-and-white video data stream with configurable parameters, namely video frame resolution and duration of the horizontal and vertical blanking intervals. The Downscaler and System Memory Model are target threads. The Downscaler downsizes the incoming image in horizontal and vertical directions by a factor of 2 (simple average between neighboring pixels is used) and attaches destination system memory addresses to the output video data stream.

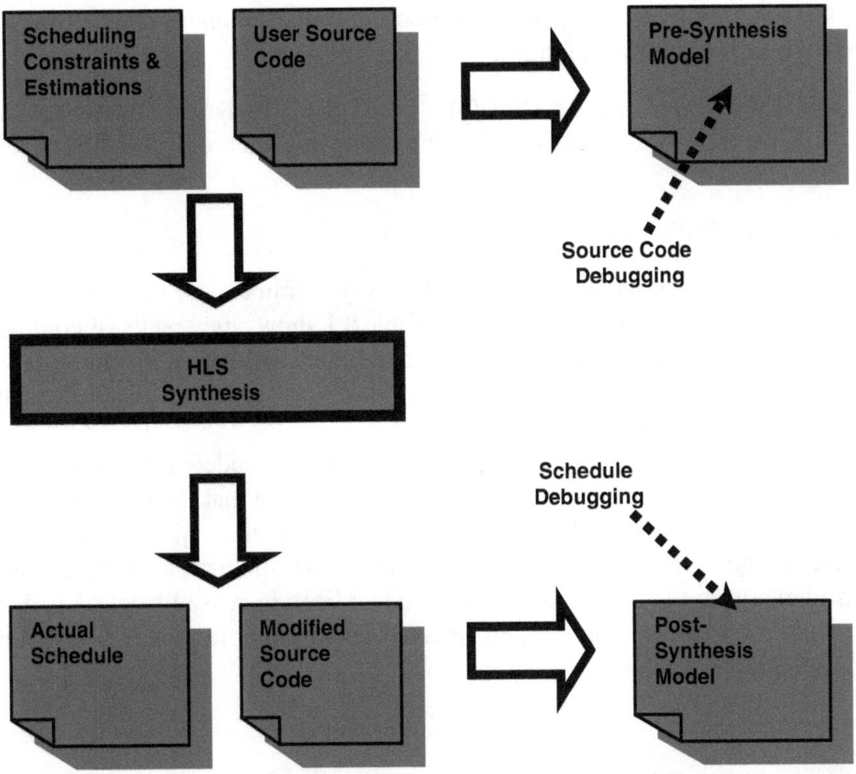

Figure 8.7. Design flow

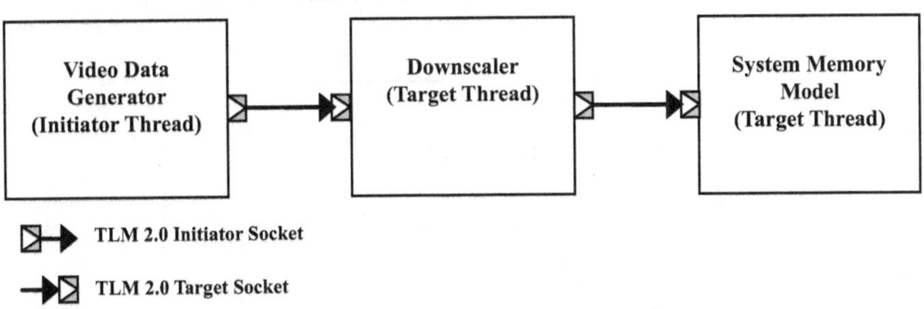

Figure 8.8. Video system used in simulation performance evaluation

For experimenting, we chose VGA video resolution (640 × 480 pixels) and assumed, for simplicity, the clock period to be equal to 10 ns. We simulated our SystemC HLS models on a machine with Intel® Core™2 Duo CPU running at

2.4 GHz and 2 Gb of RAM. We varied loosely timed simulation model parameters within the following ranges:

- Global Time Quantum (timeQ) – from 0.1 μs (re-synchronization to the SystemC clock every few pixels) to 1 ms (re-synchronization every few video lines)

- Virtual Buffer Size – from 8 to 128 pixels

We ran our simulation for 10 consecutive video frames and compared CPU time spent at each time accuracy level. Table 8.1 shows the results of comparison between the approximately timed model and the fastest (Buffer Size = 128, timeQ = 1 ms) and slowest (Buffer Size = 8, timeQ = 0.1 μs) loosely timed system models.

The results of comparison between loosely timed models of various accuracy levels are shown in Figs. 8.9 and 8.10. Please note that these results show that the most "obvious" choice of setting the buffer size equal to one video frame did not prove to be resource effective. The simulation speed did not considerably improve after the buffer size had been increased from 64 to 128 pixels because the simulation performance penalty due to context switching was already negligible.

Table 8.1. Loosely-timed vs. approximately-timed simulation performance

	Approx.-timed	Loosely-timed (slowest)	Loosely-timed (fastest)
Simulation time (s)	360	115	6

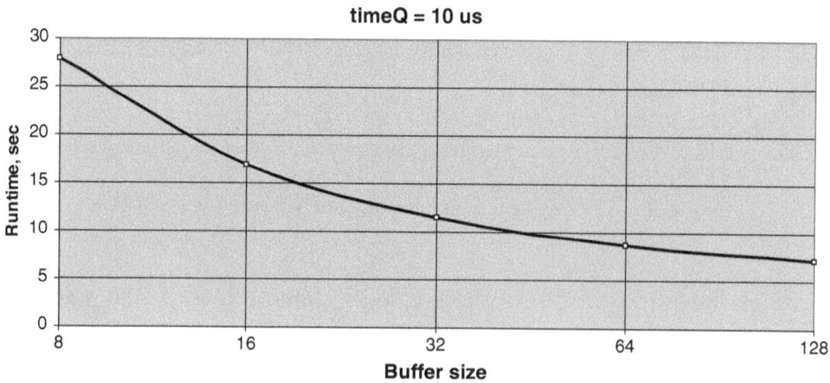

Figure 8.9. Loosely-timed simulation performance (timeQ = 10 μs)

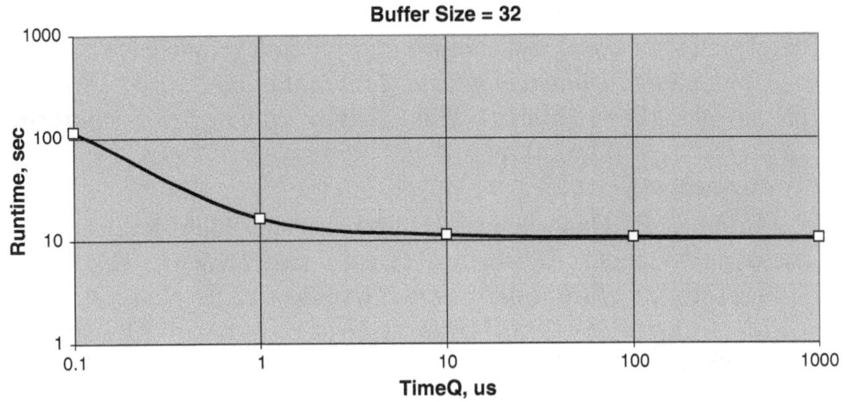

Figure 8.10. Loosely-timed simulation performance (buffer size = 32)

6. Conclusions

In this article we introduced a novel concept of Synthesizable SystemC Superset, which provides an efficient way of time-behaviour separation in HLS design specification. The main advantage and characteristic feature of our solution is a possibility of creating models that take into account the HLS scheduling process and target hardware architecture. Our solution also allows pre-synthesis and post-synthesis modeling thus greatly enhancing a designer's ability to gradually refine the system.

We also demonstrated the highly flexible level of our solution by showing that it is capable of easily switching between simulation abstraction levels without modifying the initial user specification. Our simple design example proves that by sacrificing simulation timing accuracy, simulation speed can be improved approximately 60 times, which makes our HLS specification style independent of model usage scenarios.

References

[1] R.A. Bergamaschi, S. Raje, Observable time windows: verifying high-level synthesis results. Design Test Comput. IEEE **14**(2), 40–50 (1997). URL: `http://ieeexplore.ieee.org/stamp/stamp.jsp?arnumber=587740&isnumber=12899`

[2] L.P. Carloni, K.L. McMillan, A. Saldanha, A.L. Sangiovanni-Vincentelli, A methodology for correct-by-construction latency insensitive design. Comput. Aided Design 309–315 (1999). *1999. Digest of Technical Papers. 1999 IEEE/ACM International Conference.* URL: `http://ieeexplore.ieee.org/stamp/stamp.jsp?arnumber=810667&isnumber=17544`

[3] J. Cornet, F. Maraninchi, L. Maillet-Contoz, A method for the efficient development of timed and untimed transaction-level models of systems-on-chip. *Design, Automation and Test in Europe, 2008. DATE '08*, pp. 9–14, 10–14 March 2008. URL: `http://ieeexplore.ieee.org/stamp/stamp.jsp?arnumber=4484652&isnumber=4484624`

[4] E. Grimpe, F. Oppenheimer, Object-oriented high level synthesis based on SystemC. *Electronics, Circuits and Systems, 2001. ICECS 2001. The 8th IEEE International Conference,* **1**, 529–534 (2001). URL: `http://ieeexplore.ieee.org/stamp/stamp.jsp?arnumber=957798&isnumber=20704`

[5] E. Grimpe, F. Oppenheimer, Extending the SystemC synthesis subset by object-oriented features. *Hardware/Software Codesign and System Synthesis, 2003. First IEEE/ACM/IFIP International Conference*, pp. 25–30, 1–3 Oct. 2003. URL: `http://ieeexplore.ieee.org/stamp/stamp.jsp?arnumber=1275251&isnumber=28532`

[6] IEEE Std 1666™-2005, *IEEE Standard SystemC® Language Reference Manual,* `http://standards.ieee.org`

[7] B. Niemann, C. Haubelt, Formalizing TLM with communicating state machines, in *Proceedings of Forum on Specification and Design Languages 2006 (FDL 2006)*, pp. 285–292, 2006. URL: `http://www.scientificcommons.org/21338913`

[8] Open SystemC Initiative (OSCI), `http://www.systemc.org`

[9] Open SystemC Initiative (OSCI) *OSCI TLM-2.0 User Manual*, 2007–2008, `http://www.systemc.org`; `http://ieeexplore.ieee.org/stamp/stamp.jsp?arnumber=810667&isnumber=17544`

[10] K. Pakalniškis, E. Kazanavičius, Concepts of high level synthesis using SystemC, *ISSN 1392–2114 ULTRAGARSAS, Nr.2 (43). 2002*

[11] S. Pasricha, N. Dutt, M. Ben-Romdhane, Extending the transaction level modeling approach for fast communication architecture exploration. *Design Automation Conference, 2004. Proceedings 41st,* pp. 113–118, 2004. URL: `http://ieeexplore.ieee.org/stamp/stamp.jsp?arnumber=1322453&isnumber=29281`

[12] Working Group of Open SystemC Initiative (OSCI), SystemC Synthesizable Subset, Draft 1.1.18, 2004, `http://www.systemc.org`

Part III

EMBEDDED ANALOG AND MIXED-SIGNAL SYSTEM DESIGN

Modern information processing systems frequently combine analog, RF, power electronic, or even non-electrical components with complex digital hardware and an increasing share of software into an embedded system. The aggregation and tight interaction of such components within one data processing system is a challenge: the specification, modeling, simulation, symbolic analysis, verification, design, virtual prototyping, or even synthesis of analog, mixed-signal, and heterogeneous systems, i.e., embedded systems processing both discrete and continuous signals, are highly complex issues. Furthermore, physical effects are of an increasing impact and have to be considered even at system level. Languages, models, representations, and tools such as VHDL-AMS, Verilog-AMS, SystemC-AMS, Modelica, Matlab/Simulink, Hybrid Automata or even SysML are emerging to support such issues, starting from transistor level analog circuit up to system level design. The following two chapters highlight the interaction of validation methods, modeling concepts, and related languages.

Pre-silicon verification of the compliance of a heterogeneous system with value-ranged specifications is an important task within the design flow of such systems and has to be performed by means of high-level models coded in SystemC-AMS and of sophisticated statistical methods.

RF systems design requires a consistent modeling methodology in order to address top-down/bottom-up design flows. Because of the need to consider physical effects during both architectural exploration and overall system validation, a systematic use of models coded in VHDL-AMS provides a good trade-off between simulation accuracy, fidelity, and performance.

Sorin A. Huss
FDL'09 EAMS Technical Area Chair
Technical University of Darmstadt, Germany

Chapter 9

DESIGN OF EXPERIMENTS FOR EFFECTIVE PRE-SILICON VERIFICATION OF AUTOMOTIVE ELECTRONICS

Monica Rafaila,[1] Christian Decker,[1] Christoph Grimm[2], and Georg Pelz[1]

[1] *Automotive Power, Infineon AG, 85579 Neubiberg, Germany*
monica.rafaila@infineon.com, georg.pelz@infineon.com

[2] *Vienna University of Technology, A-1040 Vienna, Austria*
christoph.grimm@ict.tuwien.ac.at

Abstract The paper presents a method to validate the compliance with value-ranged requirements of heterogeneous electronic systems with variations in a time effective way. Statistical Design of Experiments methodology is applied and adapted to plan, implement and analyze simulations of the system model, to determine key parameters that impact system performance. This is applied on a SystemC/SystemC-AMS model of an automotive electronic control unit.

Keywords: Experiment, simulation, automotive, electronic control unit

1. Introduction

The validation of heterogeneous systems against target requirements usually involves system-level model simulation. Even when the models are specified in languages like SystemC/SystemC-AMS at high abstraction levels, it can take a long time to simulate complete application scenarios that are necessary for top-level validation. When the system is subject to critical requirements, e.g. for SoCs in the automotive domain, the verification effort which would ensure

D. Borrione (ed.), *Advances in Design Methods from Modeling Languages for Embedded Systems and SoC's*, Lecture Notes in Electrical Engineering 63,
DOI 10.1007/978-90-481-9304-2_9, © Springer Science+Business Media B.V. 2010

satisfying dependability of verification is often unreasonable, as it reaches a large number of simulations. Therefore, methods are needed to permit reduction in the number of simulations that are necessary to get a certain degree of system validation. This can be realized by applying methods of statistical *Design of Experiments* (DoE), in addition to standard Monte Carlo methods, usually deployed only on circuit level.

Even if software packages e.g. [7] offer interfaces to apply classic methods for experiment planning and data analysis, they are rarely used by simulation practitioners. This particularly applies to verification of hardware/software or heterogeneous (mixed-domain, multi-nature, mixed-signal) systems, especially during architecture exploration and integration validation. Moreover, these methods must be applied on large-scale system models for fitness evaluation and potential optimization. The solution should integrate the simulator with the experiment implementation, with analysis of results and offer an interface for interactivity in this process. The simulator must support automatic model re-configuration and launching of multiple simulation runs. In this paper, it is demonstrated using the SystemC framework.

The approach is to characterize the system response, with a focus on its global maximum. Therefore, a sensitivity analysis over the input space range, focused on worst-case behavior is necessary. To identify the factors mainly responsible for the response variability, designs for simulated experiments are tested, adapted and compared. The approach is applied on a case-study of a window lifter ECU.

Specific aspects of the problem being addressed are:

- The number of simulations is limited to orders of 1 k.
- The number of input variables is high (even above 10).
- The relative variations of the input variables are small (<20%). Their statistical properties should be considered.
- The assumptions on the system response must be minimized and adapted to domain-specific particularities.

2. State of the Art

Many recent applications of DoE methods support them for input space exploration and estimation of effects. Montevechi et al. [8] exemplifies application of DoE in practical simulated experiments, but states some problems. For example, the experimental framework should be designed for simulated systems, instead of adapted from real world experimentation cases, as they differ significantly.

Modeling and simulation of electronic systems with small variations, for sensitivity and worst-case analyses, is handled in some recent approaches. For example [12] is a study of worst-case response time in distributed real-time automotive systems. Femia and Spagnuolo [1] makes worst-case analyses, with Affine Arithmetic and Genetic Algorithms, implemented on circuit simulation, while [13] proposes methods using Importance Sampling for worst-case analyses and speed-up of standard Monte Carlo.

However, existing approaches based on Affine Arithmetic lead to safe, but over-pessimistic bounds when dealing with complex effects. They offer little documentation and software implementations are hardly available. Methods based on Evolutionary and Importance sampling algorithms are considered too demanding in terms of simulation resources, for a highly dimensional verification space (require >1 M runs).

Circuit-oriented solutions are not appropriate, because at system-level, assumptions on the response need to be made and checked, to simplify the problem and reduce the number of simulations. While exploring the verification space, simulation data should be used for system characterization as well.

In this study, model specification and simulation are handled with SystemC/SystemC-AMS. For experiment planning in practice with SystemC, the authors are not aware of available implementations. Options for flexible parameter space exploration exist with SystemC Verification Library (SCV) [14]. The main disadvantages of SCV for the problem under study are that it does not provide a built-in simulation multi-run ability and that it does not support directly the generic statistic functions to represent variations in the modeled parameters.

In industrial simulation-based verification, the specification compliance must be ensured. Electronic heterogeneous systems present small variations for a multitude of parameters. Common approaches can even be limited to one of: best-guess approach (e.g. nominal value simulation); one-at-a time approach (independent analysis of each input variable); standard Monte Carlo methods; factorial methods (e.g. corner-case simulations). These can be full factorials (full exploration of the discretized input space), which are rarely used, because of the high dimension of this space. Alternatively, fractional factorial approaches would partially explore the discrete input space.

However, as [5] notes, in none of these cases a proper analysis of simulation results is made and information useful for system characterization is wasted. For specification compliance in the worst-case, pessimistic guesses on effects of input variables lead to over-estimations (oversize the specification range; over-restrict the inputs ranges). To conclude, DoE methods should be tested for applicability in planning, deploying and interpreting results of simulation-based verification flows.

3. Design of Experiments

An experiment refers to a series of tests in which changes are made to the input variables of a system to observe and identify the reasons for changes in the output, referred to hereon as response variable. Simulated experiments have the benefit of controllability over all inputs. Statistical design of experiments plans the experiment as to analyze data by statistical methods and result into objective conclusions.

Factor is a controllable experimental variable, whose variation influences the *response* variable. The values a factor can take are the levels. The *main effect* of a factor is the change in response determined by individual change of the factor, while the *interaction effect* is the failure of one factor to produce the same effect at different levels of the other factors. Both are first order effects, while a quadratic effect is a second order dependency of the response.

For experiments on simulated electronic systems, factors can be: operational conditions e.g. ambient temperature, supply characteristics, multi-nature parameters of internal/external functional blocks.

The procedure to deploy an experiment includes:

- Objective definition (e.g. characterization)
- Selection of the response variable
- Choice of factors and levels
- Choice of experimental design
- Deployment of the experiment
- Statistical and graphical analyses of the data (which can also provide with levels of confidence, thus objectivity, to statements)
- Conclusions and next steps (e.g. follow up runs)

As noted in [9], it is important to involve no more than 25% of the total resources in the first experiment.

The *worst-case* is the set of factors' values which maximizes the response. A *corner-case* is a set of factors, at the limits of their specified ranges, i.e. corners. A factor's *nominal value* is the central value of its range, while its *tolerance* is the relative distance to a corner.

Selecting an arbitrary combination of factors, i.e. best-guess approach, is frequently used in practice, based on technical knowledge of the system. One-factor-at-a-time experiment selects a starting point for each factor and successively varies each over the range. This fails to consider interaction effects. The correct approach is to conduct a *factorial* experiment, i.e. vary factors together, to study their joint effect on the response. A full factorial examines all combinations of factor levels, while a fractional factorial performs only a subset. The two-level factorial designs (i.e. sets of corner-cases) are useful in early stages to investigate main factors and their interactions. *Response surface designs* are able to estimate quadratic effects, by using more levels per factor.

The usual statistical analysis procedure starts with extracting important effects to form a preliminary *regression model*. That is, the response is approximated by a low-order polynomial and an interpolation equation is fitted by linear regression analysis. [7, 10] provide with more details on the topic. Main, interaction and quadratic effects are then coefficients of this multivariate polynomial model.

Residuals are analyzed to check the adequacy of the regression model i.e. check assumptions on important effects and on low-order empiric model for the response. Residuals are computed as differences between simulation results and predictions on the response in the simulated points. For adequate models, residuals are normally distributed with mean 0.

A normal probability plot is used to graphically assess whether sample data comes from a normal distribution, thus can evaluate adequacy of the regression model. In such a plot the y-axis is scaled such that distances on this axis match distances between the quantiles of a normal distribution (close together near the center, stretch out symmetrically as moving away). If samples are normally distributed, the plot will be linear, while other distribution types will introduce curvature. Therefore, residuals of adequate models will fall along a straight line, while samples nonlinearly distributed, or with nonzero means (i.e. center value on x-axis) indicate significant errors [7].

Graphical analysis of the regression model is also important. For example interaction plots represent the predicted response on a factor, at various levels of others. A lack of parallelism between predicted curves shows interaction with other factors. The analysis can also provide confidence bounds for each prediction.

4. Approach

Objective

The approach must characterize the response, with focus on the global maximum, over the factor multi-dimensional space. It should be flexible enough to meet various restrictions: number of runs, number of factors, assumptions on the response, valid for domain specific problems (multi-nature factors, system interdependencies). The approach must also offer systematic implementation and practical hints.

Screening and characterization experiments, as defined in [9], along system Monte Carlo methods are tested to extract factor effects, with statistical confidence associated.

The approximation with a regression model predicts the response at factor levels not yet simulated. The worst-case is estimated as the set of factors which maximizes the prediction. It remains only to check model adequacy and results will confirm that such regression models can be fitted for the simulated systems.

Abstract Description

Figure 9.1 represents the main steps in the approach.

1 The system model is validated for the specified test-case, with nominal value set for all factors. This assumes all outputs of the simulated system are checked against the expected ones, i.e. the specified ones.

2 The response variable is defined as a static characteristic of the system. It must be quantifiable and recordable after each simulation run. Its maximum is of interest to validate the worst-case against requirements.

3 Factors are set as system parameters with potential impact on the response, according to the specification, practical experience, and system

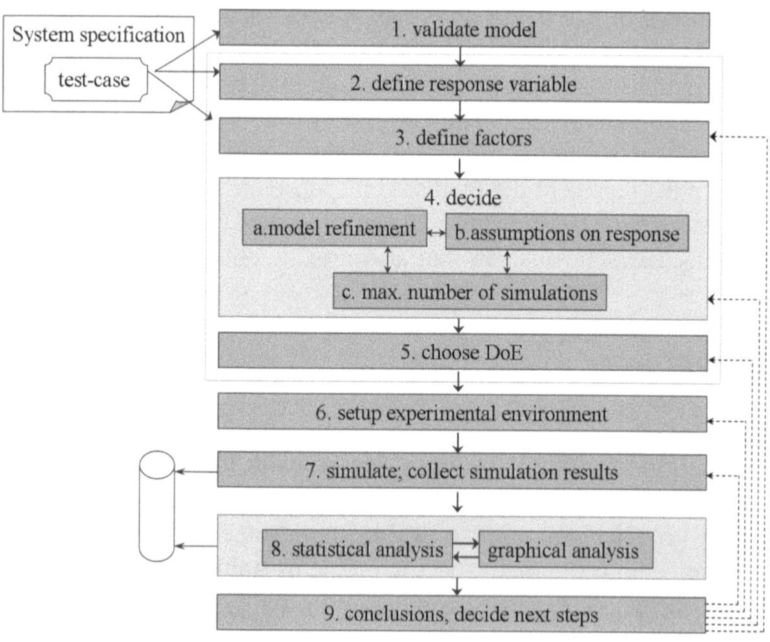

Figure 9.1. Approach steps

knowledge. The selected factors must be quantifiable, controllable and recordable. The target specification provides with allowed factor ranges and statistical properties, if existent. Such information can also be extracted from low-level simulation results or other external sources e.g. measurements. Statistically distributing the factors gives the opportunity to embed "natural" variations into models, thus validate system-level simulations.

4 (a) Choose the level of refinement for the system model. Its impact on simulation time can be observed and reduced, as long as the impact on the accuracy of results is negligible. This means equalizing, as far as possible, the simulation update and change rates of signals, with respect to the behavior in the simulated test. For this, validation against runs with higher accuracy is done, e.g. as in [11].

(b) Decide on assumptions about the response variable, or accept them as imposed by the DoE choice, made in step 5. This refers to effects assumed important, mapped to the order of the response regression model (linear – only main effects, low order – main effects and first-order interactions, second order – additional quadratic effects), or other assumptions on the response (e.g. piecewise monotonous).

(c) Decide the maximum number of simulations, and/or accept the minimum number required by the DoE choice.

For instance, a second order regression model on 10 factors requires 149 runs to be fitted. The maximum is generally restricted to orders of 1 k.

Phases a, b, c can change order, but influence each other.

5 Choose the design of the experiment. The choice depends on the maximum number of runs, on the factor choices and on the assumptions on response. Options include:

- Two-level full factorial (DoE1) includes all $2^{\#factors}$ corner-cases; regression on its results estimates all main and interaction effects.
- Multi-level full factorials (DoE2) e.g. with multiple levels for factors which present more interest (=main factors) and 2 levels for the other factors (=factor set). Effects are estimated with a special focus on the main factors.
- Response surface designs (DoE3) can estimate quadratic effects. E.g. Central Composite Design (CCD) [7, 10] adds the center point and axial runs to two-level factorials.

Fractional factorial DoEs, able to estimate main effects and some interactions, should be considered when further reduction of simulation runs is necessary.

6 Setup the experimental environment.

   This step is detailed in the Implementation section.

7 The matrix of factor levels to be simulated is loaded by the simulation controller, the experiment is then executed and results are collected.

8 The analysis of results follows the procedure detailed in the Design of Experiments Section. It must estimate as many effects as possible and predict the worst-response-case. It should also extract the overall mean and standard deviation of the response. Additional analyses depend on the DoE choice.

   Matlab toolboxes [7] are used for data handling, regression and residual analyses, and graphical studies.

9 Conclude and decide next steps (e.g. run confirmation simulations; change settings of the experimental setup; run follow-up experiments; revise the maximum number of simulations or assumptions on response; change factors).

Implementation

This section refers to setting up the model, the configuration class, the monitor and the controller (Fig. 9.2).

The model specification and basic simulation are handled with SystemC/ SystemC-AMS/ SystemC TLM [2–4]. The approach could be applied just as well with other specification languages. However, SystemC-based solutions

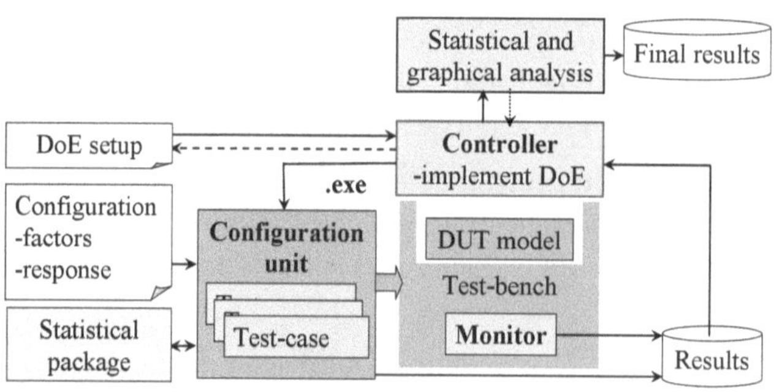

Figure 9.2. Implementation blocks

enable high expresiveness and simulation performance, in architectural level validation. Flexibility and efficiency in factor re-configuration, simulation control and response monitoring are further advantages.

The modules are parameterizable for the device under test (DUT) flexible control. Their parameters are potential factors and accessible for run-time configuration.

A *monitor* class is dedicated to record the response after each run. It can be a discrete event (DE) module (for process sensitivity e.g. when the response is a delay between events). Or it can be a timed data flow (TDF) module for sampling and evaluating of TDF signals (e.g. when the response is related to such a signal reaching a maximum or crossing a threshold). The monitor reports errors and can abort simulation. It can stop the simulation after the response has been recorded, to speed-up the experiment. It is configured with an output for report and other parameters such as thresholds to monitor.

A configuration unit is dedicated to create and configure the DUT and the factors. It acts as an interface between the simulation controller and the DUT for factor re-configuration. This class must create and configure the monitor. It must also implement the test-case (stimuli control; basic simulation handling; signal tracing; compliance checks).

The *simulation controller* mainly implements the DoE algorithm. It should handle at least:

- Offer an interface to the user e.g. to pass the number of simulations.
- Pass DoE information for DUT re-configuration.
- Launch simulation sets.
- Handle failed simulations (error or no report of monitor).
- Report status of experiment and results.

A statistical package of C++ functions, introduced in [6], is used to simulate variations in factors with statistical properties, when needed. They are set before the DUT configuration.

5. Results

System Overview

A system model of an automotive electronic control unit (ECU) designed for window lift applications is subject to experiments. The system is heterogeneous (mechanics, analog electronics, digital electronics, software) and presents multi-nature parameters with variations.

The electro-mechanical subsystem driven and supervised by the ECU has a standard structure. Its central element is a DC motor. A Hall sensor provides the ECU with the speed and direction of the DC motor, as electric signals, for window position track.

The central digital element of the ECU is the microcontroller subsystem (MCU) compatible to the standard 8051 core. A LIN transceiver is the interface for commands to control the mechanical load. Two low-side switches (LSS) are dedicated to DC motor control, by relays. A measurement block includes an interface (MI), with 8-bit ADCs, and a control and post-processing unit, which signals the MCU when error conditions occur. An operational amplifier (OPAMP) feeds the MI with amplified value of the motor current.

The ECU model uses abstract DE, TDF and TLM models of computation and communication. For the MCU, an algorithmic model and an Instruction Set Simulator were alternatively used, and validated against each other. The electro-mechanics model uses a nonlinear proprietary extension of SystemC-AMS, presently under evaluation. In this context, complete application scenarios can be simulated.

Implementation

The main functional features were verified on the system model (window control, with anti-pinch response; general protection features; power modes management).

The chosen test-case is the anti-pinch scenario (Fig. 9.3). After a command to close the window, the LSSs are controlled to drive the DC motor. An obstacle in the window determines increase in the force developed by the DC motor, thus in its current. The measurement block senses over-current and the MCU reacts, switching the LSSs off.

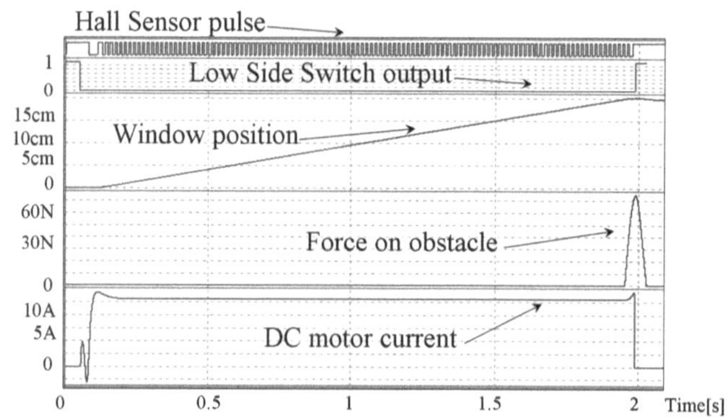

Figure 9.3. Sample simulated scenario

For this test-case, the chosen response is the maximum force sensed by the obstacle. Using the system knowledge, factors likely to impact the response variable are determined out of the specified system parameters. The choices include characteristics of the internal functional blocks involved, such as OPAMP gain (nominal 20, tolerance 1%) and LSS delay (nominal 1 ms, tolerance 10%). They can also be characteristics of external blocks – e.g. Rshunt, which is the shunt resistance used to measure the DC motor current (nominal 5 milliohms, tolerance 2%). Factors can even be stimuli, such as supply voltage Vbat (for ECU and electro-mechanical subsystem) (nominal 13.25 V, tolerance 5%).

The refinement level will set the choice for MCU model, the period of sampling TDF modules (e.g. OPAMP) and of other executions. The algorithmic MCU model is used. The TDF sample period is set to 10 ms and the MI channel evaluation period to 100 ms. The times to configure and elaborate the DUT, execute the controller and process the results are negligible compared to the simulation time.

(1) System Monte Carlo (MC)

Distributing factors according to their specified statistical properties "naturally" covers the multi-dimensional verification space. Thus the method provides an overview of the real response distribution and its statistical properties. On the other hand, it involves a large number of simulation runs.

Five hundred simulations were run, with factors normally distributed. The plot matrix in Fig. 9.4 shows scatter plots for each pair of factors, response with each factor, histograms for each factor and for the response.

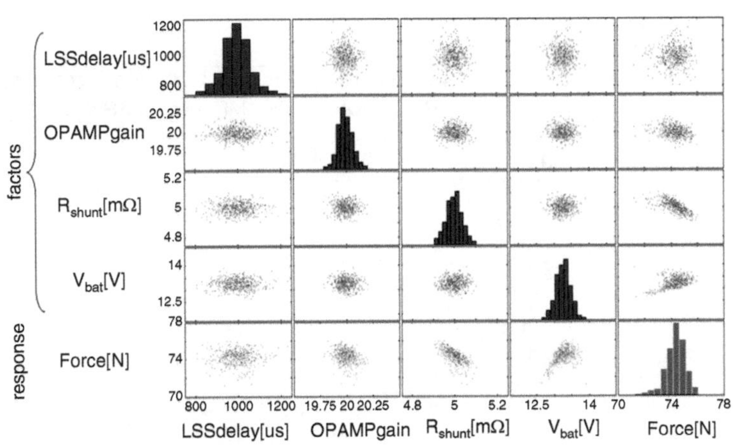

Figure 9.4. Plot matrix

As represented in histograms, factors have normal distributions: $N(\mu = \text{nominal}, \sigma^2 = (\text{tolerance}/3)^2)$.

Some interesting effects can be observed: the force increases with Vbat, for smaller values, probably because the DC motor will develop more force against the obstacle as the supply voltage increases. But for higher values, the force tends to decrease. The reason could be an ability to stop the DC motor faster and finally a smaller force. The force decreases with Rshunt and with OPAMP gain. That is because the over-current condition is triggered faster for big resistance values and higher gains. But the OPAMP gain effect is masked by more important ones. The force should increase with the LSS delay, but the effect is masked by others (the maximum variation of 200 ms in the LSS delay introduces less variability than other factors in the total reaction delay, thus in the force).

Regression analysis can make better use of the big amount of results. Coefficients of a second order regression model are estimated to quantify factor effects, and predict the worst-response-case. Response predictions with 95% confidence bounds (as dotted curves) are shown in Fig. 9.5a. The plots of the predicted response update interactively when changing the factors levels (indicated by the vertical lines), which makes the two-dimensional graphical view possible [7].

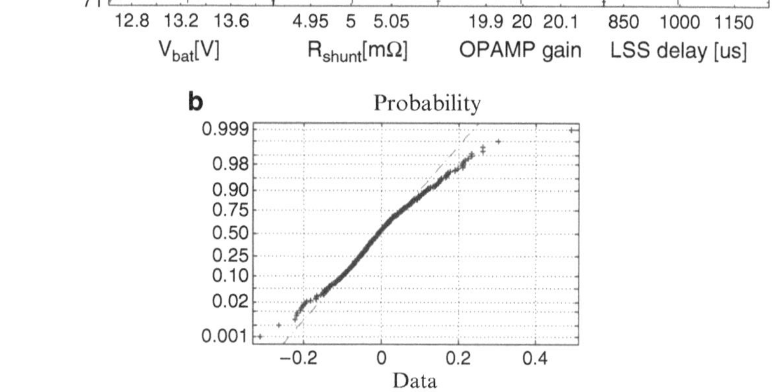

Figure 9.5. (**a**) Response prediction plots. (**b**) Normal probability plot of residuals

The residuals are computed and analyzed as described in the procedure in Section 3. Their normal probability plot (Fig. 9.5b) indicates that they are approximately normally distributed with mean 0. This confirms that the model is appropriate for this response.

Therefore, graphical analyses such as Fig. 9.4 help to identify important, even unpredictable, effects. However, a statistic analysis eliminates subjective conclusions and reduces the need of detailed system knowledge.

More important, results of a considerable number of runs indicate adequacy of regression models as general representation for major effects in such a system. However, similar conclusions can be extracted after experiments that involve fewer runs (with even first order of magnitude).

(2) Two-Level Full Factorial (DoE1)

When system Monte Carlo methods like the previous one involve too many simulation runs, the two-level full factorial DoE ($2^{\#factors}$ corners) can be implemented. Similar analysis, but on much less data, is performed and the fitted regression model is less adequate.

This DoE is able to estimate only main and interaction effects. The residuals were represented on a normal probability plot (Fig. 9.6b) and indicate lack of fit, as they do not lie along a straight line. Therefore, the assumption of no quadratic effects in response does not hold.

The results can still be used, e.g. to observe interaction effects and/or potential sources of inadequacy in the meta-model. Figure 9.6a represents the estimated regression model. Potential interaction effects are visible in interactive plots: the predicted response values and confidence intervals update when modifying the factor levels [7]. E.g. an interaction between OPAMP gain and Vbat is visible.

Still, the residuals are not close enough to a normal distribution with mean 0 and the confidence bounds are rather large. A follow-up is necessary, such as DoE2.

(3) Multi-level Full Factorial (DoE2)

DoE1 results pointed Vbat as an interesting factor, due to interactions with OPAMP gain and Rshunt.

DoE2 is a custom implementation of a multi-level full factorial, with a special focus on the main factor, and corresponding analysis of results and estimation of effects: Vbat is the main factor, with 10 equidistant levels spread across its range. Simulation sets, with the other factors (=factor set) set to corners, are run at each main factor level. Alternatively, the factor set could be statistically distributed. Results show an interaction between the main factor and the factor set, plus a non-monotonous response, over the main factor range.

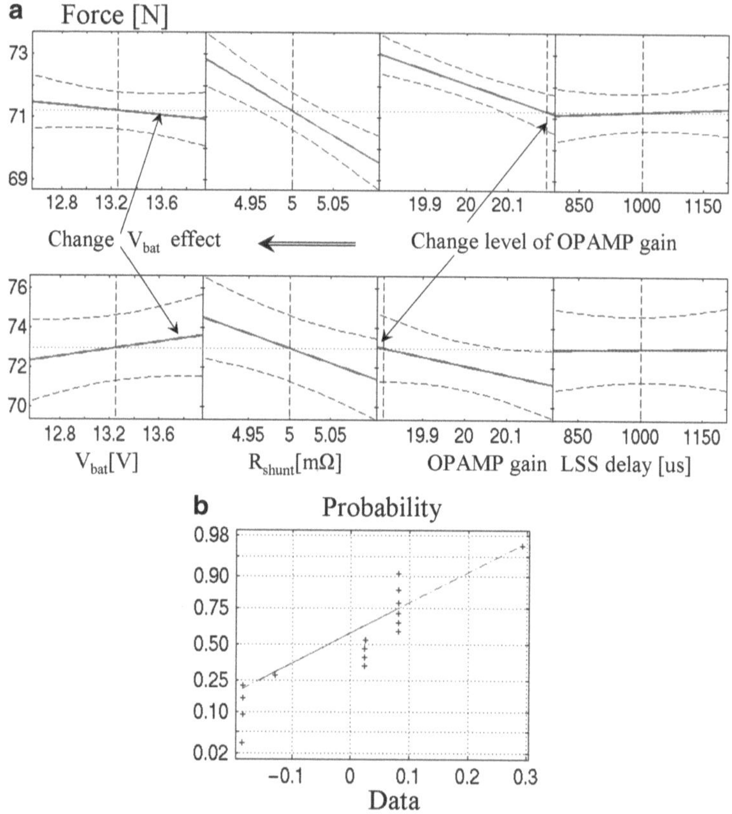

Figure 9.6. (**a**) Response prediction plot. (**b**) Normal probability plot of residuals

The force is interpolated over main factor levels, thus assumed piecewise linear, for each set. Figure 9.7 shows that the worst-case is not a main factor corner, and that interaction exists (plots of different factor sets are not parallel). Furthermore, the worst-case corresponds to different levels of the main factor, depending on the choice for factor set corners. The response variability on Vbat is different for different sets, i.e. another sign of interaction.

The change in response over main factor levels, relative to the distance between different sets, shows the ratio of main effects (main factor/factor set). The contrast between deviations within sets at different main factor levels measures the interaction main factor – factor set. Estimates are summarized in Table 9.1, if DoE2 is run for each of the factors.

Similar conclusions can be drawn based on the interpolated Force (factor set index).

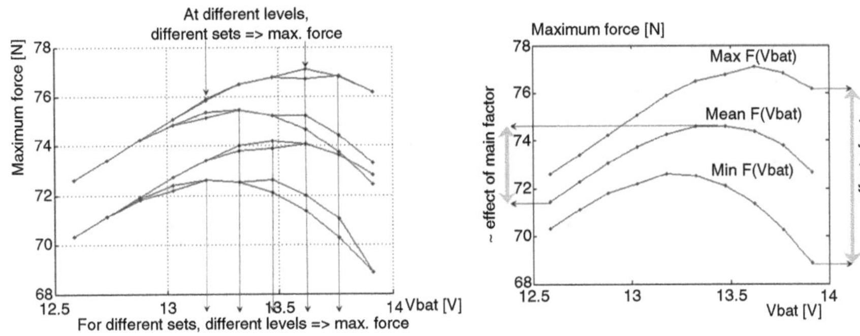

Figure 9.7. Interpolated response

Table 9.1. Estimates of effects

Effect estimates [N]		Vbat	Rshunt	OPAMP gain	LSS delay
Main factor effect	mean (std_sets)	1.2727	1.0006	0.6171	0.0556
Factor set effect	std (mean_sets)	1.6235	1.5059	1.7703	2.2428
Interaction	std (std_levels)	0.5323	0.0609	0.2252	0.0219

(4) Response Surface DoE – CCD (DoE3)

Running all corners (2^4 runs), the center point and two axial runs for each factor yields, after regression, a full quadratic regression model. Thus this experiment is able to locate the second order effect of Vbat and other important interactions. Quantifying these effects allows comparisons and prediction of the worst-response-case. The interactive prediction plots are represented in Fig. 9.8a.

The predicted response is:
$\text{Force} = 75.71 + 1.04 \cdot F_{1n} - 1.44 \cdot F_{2n} - 0.67 \cdot F_{3n} + 0.11 \cdot F_{4n} - 0.33 \cdot F_{1n} \cdot F_{2n} - 0.75 \cdot F_{1n} \cdot F_{3n} + 0.12 \cdot F_{1n} \cdot F_{4n} - 0.07 \cdot F_{2n} \cdot F_{3n} - 0.04 \cdot F_{2n} \cdot F_{4n} - 0.03 \cdot F_{3n} \cdot F_{4n} - 2.04 \cdot F_{1n}^2 + 0.01 \cdot F_{2n}^2 - 0.09 \cdot F_{3n}^2 - 0.01 \cdot F_{4n}^2 [N]$
where $F_{kn} = F_k$/(nominal value of factor).

Residuals were found distributed approximately linearly on the normal probability plot with zero means (Fig. 9.8b), which confirms the model adequacy.

Estimated regression coefficients were also validated against the ones extracted after the Monte Carlo experiment (which involved many runs, thus provides more statistical confidence to the estimated effects).

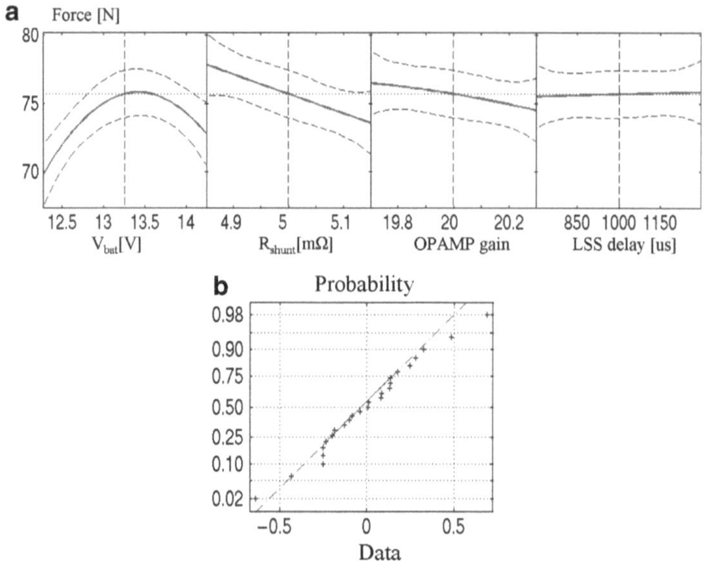

Figure 9.8. (**a**) Response prediction plot. (**b**) Normal probability plot of residuals

Discussion

The worst-cases (predicted and/or simulated) and the number of runs for tested experiments are in the following order:

WC3(78.16N) > WC2(77.09N) > MC(76.41N) > WC1(76.18N)
No.Sim.MC(500) > No.Sim.2(80) > No.Sim.3(25) > No.Sim.1(16)
Their overall response mean and variability were:

- MC: Mean = 74.3010N; Std. dev. = 0.9291N
- DoE1: Mean = 72.07N; Std. dev. = 2.17N
- DoE2: Mean = 73.49N; Std. dev. = 1.96N
- DoE3: Mean = 73.99N; Std. dev. = 2.22N

A comparison of experiments is summarized in Table 9.2.

For DoE1, the assumption of no quadratic effects limits the model adequacy, thus the quality of results. Unlike DoE1, DoE2 and DoE3 are able to locate cases worse than worst corner-case, when higher-order effects are present, while MC and DoE2 can simulate statistical variations. DoE2 provides efficient analysis of the main factor, but it offers no clear prediction with respect to the factor set and no adequacy check. It can only predict based on response interpolation, assumed piecewise linear.

Table 9.2. DoE comparative analysis

DoE	No. sim. runs	Response prediction	Effects estimates	Graphical representation	Statistical factor properties	Implementation effort
MC	–	–	–	++	++	+
DoE1	+++	+	+	+	–	++
DoE2	+	+	++	++	+	+++
DoE3	++	++	+++	++	–	++

DoE3 performs best for response approximation and effect estimates. It is the best trade-off between number of runs and quality of system characterization, with worst-case prediction.

6. Conclusion

Proper planning and analysis of simulation experiments can improve characterization and validation of electronic systems. This was applied on a heterogeneous system with small factor variations, modeled with SystemC/ SystemC-AMS.

The statistical analysis estimates factor effects, with associated confidence, and predicts the response, thus the worst-case. Steps of the implementation enable flexibility in testing designed experiments, either classic or adapted.

The number of simulation runs increases with the DoE efficiency, given by the regression model adequacy and by the quality worst-case estimation.

Monte Carlo methods can and should be involved in preliminary phases, to visualize main effects. They enable embedding of low-level information into high-level, fast models, about statistic distributions of factors and their inter-dependencies (e.g. generated by common sources of variability).

Further work is focused on assessing the applicability to the present problems of other designed experiments, with high confidence in predicting the response and low requirements for the number of simulation runs. Adaptive experimentation and optimization algorithms are also under evaluation.

Acknowledgement

This research project is supported by the German Government, Federal Ministry of Education and Research under the grant number 01M3178*. The authors are responsible for the context of the paper.

References

[1] N. Femia, G. Spagnuolo, True worst-case circuit tolerance analysis using Genetic Algorithms and Affine Arithmetic, Circuits and Systems I: Fundamental Theory and Applications, 2000, pp. 1285–1296

[2] F. Ghenassia, *Transaction-Level Modeling with SystemC: TLM Concepts and Applications for Embedded Systems*, Springer-Verlag New York, Inc., Secaucus, NJ, USA, 2006, ISBN: 0387262326

[3] C. Grimm, M. Barnasconi, A. Vachoux, K. Einwich, An introduction to modeling Embedded Analog/Mixed-Signal Systems using SystemC AMS extensions, http://www.systemc.org/downloads/, 2008

[4] T. Grötker, S. Liao, G. Martin, S. Swan, *System Design with SystemC* (Kluwer, 2002)

[5] A. Law, Statistical analysis of simulation output data: The practical state of the art, WSC '04: Proceedings of the 36th conference on Winter simulation, Washington D.C., 2004, pp. 67–72, ISBN: 0-7803-8786-4

[6] T. Markwirth, J. Haase, K. Einwich, Statistical Modeling with SystemC-AMS for automotive systems, in *Forum on Specification and Design Languages*, 2008, pp. 247–248

[7] Matlab, www.mathworks.com/access/helpdesk/help/toolbox/stats.

[8] J.A.B. Montevechi, A.F. de Pinho, F. Leal, F.A.S. Marins, Application of design of experiments on the simulation of a process in automotive industry, in *Winter Simulation Conference*, 2007, pp. 1601–1609

[9] D.C. Montgomery, *Design and Analysis of Experiments*, John Wiley & Sons, 2006, pp. 1–22, ISBN: 0470088109

[10] D.C. Montgomery, *Design and Analysis of Experiments*, John Wiley & Sons, 2006, pp. 203–250, ISBN: 0470088109

[11] M. Rafaila, C. Decker, C. Grimm, K. Einwich, T. Markwirth, G. Pelz, New methods for system-level verification using SystemC-AMS extensions: application to an automotive ECU, in *Proceedings of the Methoden und Beschreibungssprachen zur Modellierung und Verifikation von Schaltungen und Systemen*, Berlin, 2009

[12] S. Samii, S. Rafiliu, P. Eles, P. Zebo, A simulation methodology for worst-case response time estimation of distributed real-time systems, Design, Automation and Test in Europe, 2008, pp. 556–561

[13] A. Singhee, R.A. Rutenbar, Statistical blockade: a novel method for very fast Monte Carlo simulation of rare circuit events and its application, Design, Automation and Test in Europe, 2007, pp. 1–6

[14] SystemC Verification Standard Specification Version 1.0e, `http://www.systemc.org/downloads/standards/2003`

Chapter 10

A VHDL-AMS MODELING METHODOLOGY FOR TOP-DOWN/BOTTOM-UP DESIGN OF RF SYSTEMS

Torsten Maehne,[1] Alain Vachoux,[1] Frédéric Giroud,[2] and Matteo Contaldo[2]

[1] *Laboratoire de Systèmes Microélectroniques (LSM)*
École Polytechnique Fédérale de Lausanne (EPFL)
EPFL/STI/IEL/LSM, Bâtiment ELD, Station 11, CH-1015 Lausanne, Switzerland
WWW: http://lsm.epfl.ch/
torsten.maehne@epfl.ch
alain.vachoux@epfl.ch

[2] *Microelectronics Division*
Centre Suisse d'Électronique et de Microtechnique SA (CSEM)
Rue Jaquet-Droz 1, P.O. Box, CH-2002 Neuchâtel, Switzerland
WWW: http://www.csem.ch/
frederic.giroud@csem.ch
matteo.contaldo@csem.ch

Abstract This chapter presents a modeling methodology for the top-down/bottom-up design of RF systems based on the systematic use of VHDL-AMS models. The model interfaces are parameterizable and pin-accurate. The designer can choose to parameterize the models using performance specifications or device parameters back-annotated from the transistor-level implementation. The abstraction level used for the description of the respective analog/digital component behavior has been chosen to achieve a good trade-off between accuracy, fidelity, and simulation performance. These properties make the models suitable for different design tasks such as architectural exploration or overall system validation. This is demonstrated on a model of a binary FSK transmitter parameterized to meet very different target specifications. The achieved flexibility and systematic model documentation facilitate their reuse in other design projects.

Keywords: RF systems, binary FSK transmitter, modeling methodology, `RF_TRX` model library, VHDL-AMS

D. Borrione (ed.), *Advances in Design Methods from Modeling Languages for Embedded Systems and SoC's*, Lecture Notes in Electrical Engineering 63,
DOI 10.1007/978-90-481-9304-2_10, © Springer Science+Business Media B.V. 2010

1. Introduction

Portable, battery-powered electronic devices (cell phones, PDAs, notebooks, mice, ...) nowadays routinely contain wireless functionality to communicate among each other using standards like Bluetooth™, Wi-Fi™, or GSM™. This became possible due to the integration of full RF transceivers into a System-on-a-Chip (SoC), which are built of these main parts (Fig. 10.1):

- **Frequency Synthesizer:** Based on a Phase-Locked Loop (PLL) consisting of a Voltage Controlled Oscillator (VCO), DIViders (DIVs), Phase/Frequency Detector (PFD), Charge Pump (CP), LooP Filter (LPF), Quartz Oscillator (QO) for the reference frequency, Σ-Δ modulator for fractional frequency division and Multi-stAge-noise-SHaping (MASH) to generate the RF carrier signal required for demodulation and modulation.

- **Receiver (RX) Chain:** Channel Selector (CS), Low-Noise Amplifier (LNA), MIXer (MIX), Intermediate Frequency Amplifier (IFA), Poly-Phase Filter (PPF), and Digital Signal Processing (DSP).

- **Transmitter (TX) Part:** Modulation of the Pulse-Shaped (PS) data bit stream on the RF carrier using direct modulation of the fractional PLL and its amplification with a Power Amplifier (PA).

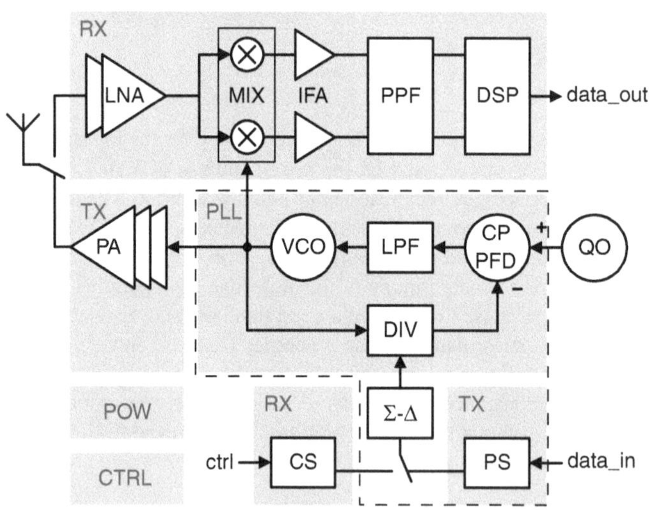

Figure 10.1. A global view of the RF transceiver

- **POWer Management (POW):** Linear regulators, step-up converters.
- **Digital ConTRoL (CTRL)** of the transceiver through some communication interface (e.g., SPI or I^2C bus) or any other digital system (e.g., microcontroller).

The design of such a complex system is highly demanding since its analog/RF and digital parts require different design flows, methodologies, and tools.

During the *top-down design phase*, the overall system specifications must be properly distributed and assigned to the components defined by the selected architecture. Then, the result has to be evaluated and checked against constraints such as the feasibility with the selected technological process. The distributed specifications will then be used for the detailed block design until their layout is obtained. While Hardware Description Languages (HDLs) such as VHDL or Verilog are routinely used for the design of digital parts, mathematical tools such as MATLAB/Simulink or spreadsheets are preferred for deriving analog/RF specifications and performing system-level studies. However, mixed-signal HDLs such as VHDL-AMS or Verilog-AMS do offer additional capabilities that can support the top-down design and more specifically the architectural exploration step. Anticipating the real component behaviors with models at proper levels of abstraction using appropriate parameters is a considerable help in that phase. Their reusability for evaluating various possible architectures is also an advantage [4].

During the *bottom-up verification phase*, the implemented components must be individually verified and then assembled to form the complete system under design. Assuming that each component has been successfully verified, a full top-level verification is necessary for detecting interface issues such as wrong signal behavior or incorrect timing. As full system verification at transistor level is becoming impracticable due to the huge amount of data to handle, models at proper levels of abstraction using appropriate parameters again offer the required accuracy/verification performance trade-off [11].

This chapter reports on a work that has the goal to develop and provide a library of VHDL-AMS models called `RF_TRX` that supports both the top-down architectural exploration and the bottom-up verification of complex RF systems such as the RF transceiver. VHDL-AMS models of components of a Frequency Shift Keying (FSK) transmitter (indicated by the dashed line in Fig. 10.1) have been developed following a particular modeling methodology, as presented in Section 2. Special care has been taken to model the proper non-idealities, while conserving acceptable simulation performances, and to define consistent parameter sets that support the top-down (specification) and bottom-up (back-annotation) phases. Similar model libraries have been already presented, e.g., in [3, 8, 9], but they mostly focus on top-down design only.

The chapter is organized as follows. Section 2 presents the modeling methodology that has been used for the PLL in the frequency synthesizer. The development of a model of the VCO is described in detail, from the designer's specification to the actual VHDL-AMS model. Other component models needed to model the frequency synthesizer are only outlined as the space is limited. Section 3 presents the usage of the developed VHDL-AMS models for the architectural exploration of an FSK transmitter in two different design cases using carrier frequencies of 868 MHz and 2.45 GHz, how the developed structural transmitter model can be reused during bottom-up verification, and how modeling decisions impact the simulation performance. Finally, Section 4 includes some conclusions and an outlook on future steps.

2. Modeling Methodology

In this section, we outline the modeling methodology and illustrate its application using the frequency synthesizer and particularly its VCO component as an example.

The first step involves the *definition of the requirements the model library shall meet.* This includes the definition of the components that will be modeled, the level of detail, at which the component behaviors shall be described, the parameters that shall be available to the model user, and the parameters that shall be derived internally. The definition of these two kinds of parameters has to be done carefully as the same model shall be used for both top-down architectural exploration and bottom-up verification. In the top-down phase, the interface (generic) parameters are used as specifications and internal parameters are used as nominal values for the design of the component at the transistor level. In the bottom-up phase, the interface parameters are back-annotated with values from the transistor-level implementation and internal parameters have in turn values that correspond to extracted specifications. This also means that models shall be pin-accurate for easy replacement with equivalent transistor-level models, when required. The second step involves the *specification of the behaviors to model, their interfaces (signals and parameters), and their internal parameters* as discussed above. This depends largely on which specifications need to be validated with the help of the (sub-)system model and requires a preliminary understanding of the overall system function to derive the principal component parameters that have an influence on the system performance. The specification is largely prepared by the domain experts, i.e., in our case the RF designers, with feedback given by the model developers, e.g., asking for clarification or restructuring of certain specifications. Then, the *agreed specifications have to be translated into legal VHDL-AMS models.* In this step, the capabilities offered by the modeling language and, unfortunately, the incomplete support of the language in existing EDA tools, have to be balanced to achieve the intended accuracy/fidelity and simulation performance. Finally, the *models shall be thoroughly tested using elaborate test benches* that include as much as possible the verification of assertions about the modeled components.

As these steps imply a non-negligible implementation effort, the models shall be as flexible/modular as possible to be *reusable in other design projects* to accelerate and optimize the design process using an over time growing base of well documented and individually reviewed and verified component models, in which validity other designers can trust. The particular requirements for such a model library are outside the scope of this chapter, but are described in more detail in [6].

Specifications for the Frequency Synthesizer

A frequency synthesizer [12] is a system that allows to enslave an oscillator (VCO) at a specific frequency. It is most of the time based on a PLL (Fig. 10.1). For the RF application described here, the VCO is designed to give a signal at high frequencies, like those specified for the ISM or SRD bands. The frequency of the VCO will change until its frequency, divided by the ratios of the frequency dividers, is exactly equal to the frequency of the quartz oscillator. The equation expressing the relation between VCO frequency, the division ratio, and the quartz reference frequency is:

$$f_{\text{vco}} = f_{\text{REF,PLL}} \cdot N_{\text{div}} \tag{10.1}$$

Variables with uppercase subscripts will be considered later as model constants (either generic parameters or local constants) and variables with lowercase subscripts will be considered later as model quantities or signals.

As for every looped system, a *stability problem* occurs if the phase margin of the PLL open-loop gain is too small. For calculating this open-loop gain, we need to define the principal parameters (Fig. 10.2) of each block:

- **VCO:** K_{vco} in Hz/V, expresses the relationship between the input voltage v_{tune} and the output VCO frequency.
- **Frequency Divider:** N_{div}, modulated by the Σ-Δ modulator, is the division ratio between the VCO and the quartz oscillator frequencies. In an implementation N_{div} might be obtained by cascading several dividers.
- **PFD and CP:** K_{PD} in A/rad, expresses the gain between the phase error at the input of the PFD and the instantaneous current flowing at the output of the CP.
- **LPF:** C_1, R_2, C_2, R_3, C_3 are the circuit device parameters of the loop filter, which need to be adjusted to get a good phase margin Φ_{MARGIN} for a specific cut-off frequency $f_{\text{CUT-OFF}}$. Note that a fourth order R_4, C_4 (impedance at the VCO input) may be taken into account.

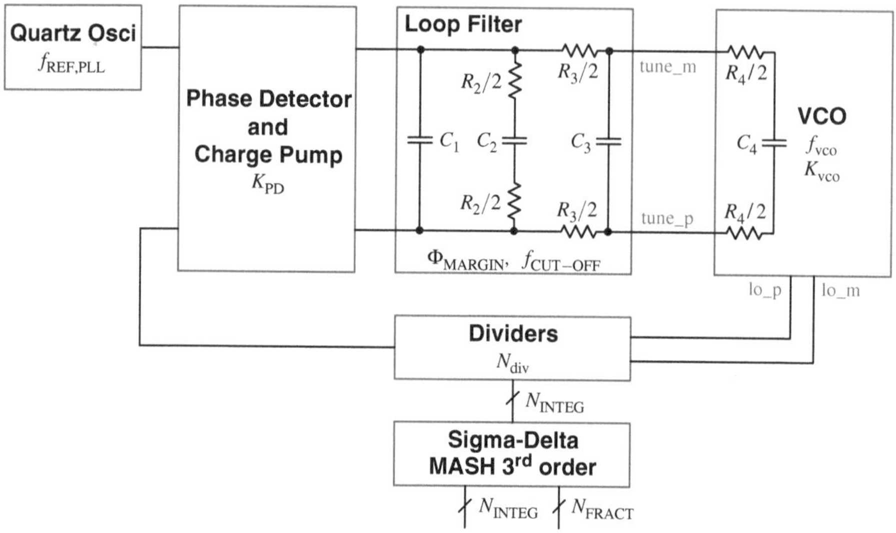

Figure 10.2. A global view of the frequency synthesizer

Using these parameters, the open-loop transfer function of the PLL is given by the following equation:

$$H(j\omega) = \frac{\frac{K_{\mathrm{vco}} \cdot K_{\mathrm{PD}}}{N_{\mathrm{DIV}} \cdot j\omega} \cdot Z(j\omega)}{\left(1 + j\omega \cdot R_3 \cdot \left(C_3 + \frac{C_4}{1+j\omega \cdot R_4 C_4}\right)\right) \cdot (1 + j\omega \cdot R_4 C_4)} \tag{10.2}$$

where $Z(j\omega)$ is the impedance seen by the charge-pump:

$$Z(j\omega) = \frac{1}{j\omega \cdot \left(C_1 + \frac{C_2}{1+j\omega \cdot R_2 C_2} + \frac{C_3 + \frac{C_4}{1+j\omega \cdot R_4 C_4}}{1+j\omega \cdot R_3 \cdot \left(C_3 + \frac{C_4}{1+j\omega \cdot R_4 C_4}\right)}\right)} \tag{10.3}$$

By calculating the phase value, for which the magnitude of Eq. 10.2 is equal to one, the *system phase margin* can be derived [10].

Different implementations can be used to build the *VCO* by satisfying the gain and phase conditions needed to ensure steady-state oscillation [5]. When used in a frequency synthesizer, the oscillator has to cover a precise output frequency range determined by the target application. A tunable ideal oscillator with continuous output phase and control voltage can be represented by the following equation:

$$v_{\mathrm{osc}}(t) = V_{\mathrm{AMP}} \cdot \cos\left(2\pi \left(K_{\mathrm{vco}} \int_{-\infty}^{t} v_{\mathrm{tune,eff}}\, dt + f_{\mathrm{vco,0}} \cdot t\right)\right) \tag{10.4}$$

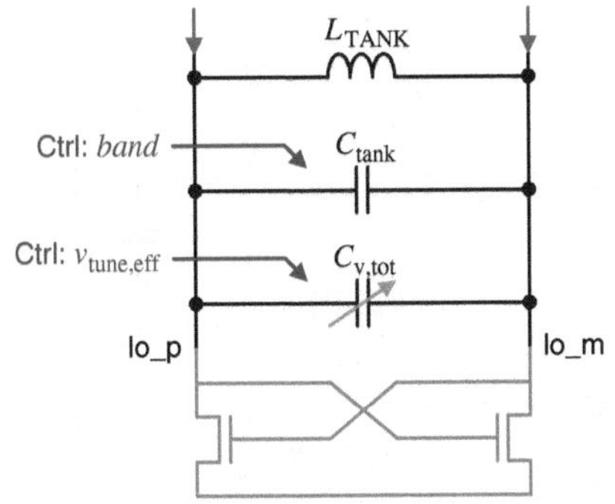

Figure 10.3. Resonant elements of the VCO

The oscillator taken into account for the modeling is an LC oscillator (Fig. 10.3) with band switching capabilities. It is composed by an inductor and an equivalent capacitor needed to fix the oscillator frequency, while some active devices compensate for the losses in order to sustain the oscillation.

The frequency tuning is performed by adjusting the capacitance value using an analog and a digital control in parallel. The analog control is achieved by means of a varicap allowing continuous frequency tuning over a certain range. The digital control is realized as a bank of switchable capacitances, which allows to shift the center frequency $f_{\text{vco},0}$ of the range covered with the analog control in a discrete way. The final oscillation frequency is thus given by:

$$f_{\text{vco}} = \frac{1}{2\pi \cdot \sqrt{L_{\text{TANK}} \cdot (C_{\text{tank}}(band) + C_{\text{v,tot}}(v_{\text{tune,eff}}))}} \tag{10.5}$$

with L_{TANK} the oscillator inductance, $C_{\text{tank}}(band)$ the bank capacitance for the selected $band$, $C_{\text{v,tot}}(v_{\text{tune,eff}})$ the variable capacitance for the effective tuning voltage $v_{\text{tune,eff}}$.

The previous equations show that the primary parameter needed for a correct VCO model is K_{vco}. It is present in the time- and frequency-domain equations (by means of the variable capacitance) affecting not only the VCO behavior but also the PLL stability (Eq. 10.2). The value of K_{vco} is directly related to the characteristic of the varicap, which is the circuit element used to perform the analog frequency tuning. Therefore, a detailed study of the varicap and its usage in the VCO circuit has to be done. Fig. 10.4 depicts the varicap configuration inside the VCO structure with biasing resistances R_4, coupling

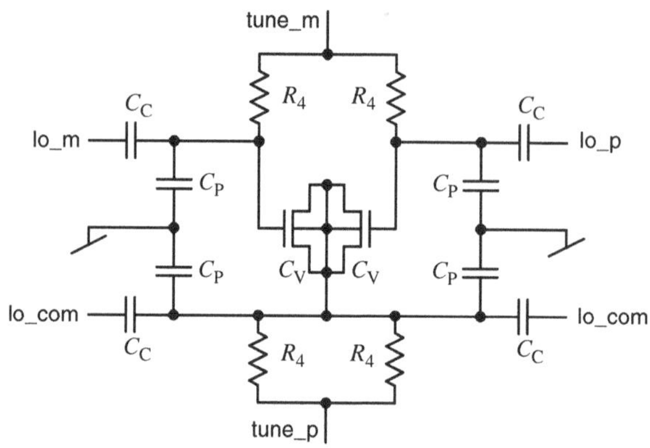

Figure 10.4. Wiring of the varicap inside the VCO structure

capacitances C_C, and parasitic capacitances C_P. The equivalent capacitances seen from the filter and from the oscillator terminals are:

$$C_4 = 2C_{\mathrm{v}} + C_{\mathrm{C}} + C_{\mathrm{P}} \tag{10.6}$$

$$C_{\mathrm{v,tot}} = \left(\frac{2}{C_{\mathrm{C}}} + \frac{2}{C_{\mathrm{v}} + C_{\mathrm{P}}}\right)^{-1} \quad \text{with} \quad \frac{C_{\mathrm{P}}}{C_{\mathrm{C}}} \cong 1\,\% \tag{10.7}$$

The nonlinear dependency of the varicap density from the effective tuning voltage $v_{\mathrm{tune,eff}}$ at its terminals can be accurately described using a hyperbolic tangent function:

$$D_{\mathrm{cv}}(v_{\mathrm{tune,eff}}) = D_{\mathrm{CV,AMP}} \cdot \tanh\left(\frac{\frac{dD_{\mathrm{CV},0}}{dv_{\mathrm{tune,eff}}}}{D_{\mathrm{CV,AMP}}} \cdot (v_{\mathrm{tune,eff}} - V_0)\right) + D_{\mathrm{CV},0} \tag{10.8}$$

with the following technology-dependent parameters (Fig. 10.5):

- V_0: Offset tuning voltage for the point with the steepest slope $\frac{dD_{\mathrm{cv}}}{dv_{\mathrm{tune,eff}}}$ of the measured varicap density.
- $D_{\mathrm{CV},0} = D_{\mathrm{cv}}(V_0)$: Varicap density measured for V_0.
- $D_{\mathrm{CV,AMP}}$: Varicap density modulation amplitude.
- $\frac{dD_{\mathrm{CV},0}}{dv_{\mathrm{tune,eff}}} = \left.\frac{dD_{\mathrm{CV}}}{dv_{\mathrm{tune,eff}}}\right|_{V_0}$: Slope of D_{cv} measured at V_0.

By deriving the varicap density function with respect to the tuning voltage $v_{\mathrm{tune,eff}}$, we obtain:

$$\frac{dD_{\mathrm{cv}}}{dv_{\mathrm{tune,eff}}} = \frac{dD_{\mathrm{CV},0}}{dv_{\mathrm{tune,eff}}} \cdot \left(1 - \left(\frac{D_{\mathrm{cv}}(v_{\mathrm{tune,eff}}) - D_{\mathrm{CV},0}}{D_{\mathrm{CV,AMP}}}\right)^2\right) \tag{10.9}$$

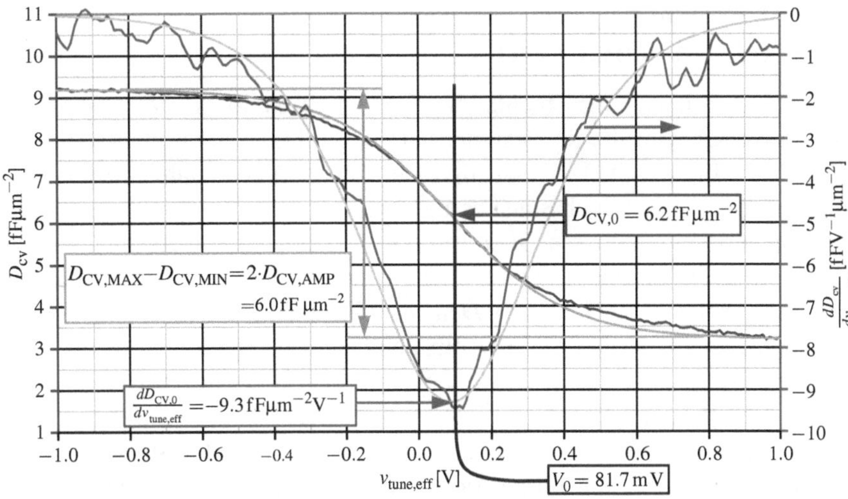

Figure 10.5. Fitting of varicap density and its derivative for TSMC 0.18 µm

Then, by defining the area A_{CV} of the varicap, it is possible to calculate its absolute value and derivative:

$$C_v = A_{CV} \cdot D_{cv} \qquad \frac{dC_v}{dv_{tune,eff}} = A_{CV} \cdot \frac{dD_{cv}}{dv_{tune,eff}} \tag{10.10}$$

To calculate K_{vco}, we still need the derivative $\frac{dC_{v,tot}}{dv_{tune,eff}}$ of the total equivalent varicap $C_{v,tot}$ seen by the VCO between the terminals lo_p and lo_m taking into account the coupling and parasitic capacitances present in the circuit (Fig. 10.4):

$$\frac{dC_{v,tot}}{dv_{tune,eff}} = \frac{1}{2} \cdot \frac{1}{\left(1 + \frac{C_v}{C_C} + \frac{C_P}{C_C}\right)^2} \cdot \frac{dC_v}{dv_{tune,eff}} \tag{10.11}$$

Using Eqs. 10.5 and 10.11, we can finally calculate K_{vco}:

$$K_{vco} = \frac{df_{vco}}{dv_{tune,eff}} = 2\pi^2 \cdot f_{vco}^3 \cdot L_{TANK} \cdot \left(-\frac{dC_{v,tot}}{dv_{tune,eff}}\right) \tag{10.12}$$

Using the technology constants introduced for the varicap density given by Eq. 10.8 and choosing the ratios $\frac{C_C}{C_{V,0}}$ and $\frac{C_P}{C_C}$ for the capacitances, the varicap area A_{CV} and its related capacitances C_C and C_P can be calculated to optimize the VCO to have a maximal K_{vco} at a certain target frequency $f_{VCO,OPT}$. This optimal frequency is supposed to be at the inflection point of $D_{cv}(v_{tune,eff})$ at V_0. This maximal K_{vco} is used for the PLL stability calculation with the help of Eqs. 10.2 and 10.3. For accurate simulation, the instantaneous value of K_{vco} is always used.

Design, Implementation, and Test of the Frequency Synthesizer

During the design and implementation phase, the functional specification developed together with the RF designer is translated into an executable model. In the RF_TRX library, the PFD, CP, dividers, and Σ-Δ modulator models have a single architecture, which satisfies the needs for the top-down design and bottom-up verification phases. Their top-down design parameters are also used during the bottom-up phase, which adds just additional details like delays or mismatches. The *Phase/Frequency Detector (PFD) model* is implemented as a digital behavior description (two D-flip-flops and one AND-gate) augmented by generics to specify the propagation delays of the gates back-annotated during the bottom-up phase. The *differential charge pump model* is implemented as ideally switched current sources with internal resistance. Their value is determined as a multiple of a measured input bias current, as they are implemented as current-mirrors on transistor level. Mismatches can be specified for the bottom-up verification. This level of precision is sufficient as, for a more detailed study of the charge pump's impact (especially the switching effects from the transistors) on the PLL performance, many simulators allow the mixed simulation of VHDL-AMS and SPICE component models.

The implementation of the *differential loop filter model* follows the stability analysis of the PLL. It models the RC network as branch equations. This allows to use simultaneous **if**-statements to implement the filter orders 1, 2, and 3 in the same model. For the top-down design phase, the RF designer can specify in the generics, besides the filter cut-off frequency $f_{\mathrm{CUT-OFF}}$ and order, the phase margin Φ_{MARGIN} to be realized for the PLL characterized by the charge pump sensibility K_{PD}, the VCO frequency sensibility K_{VCO}, and the frequency division factor N_{DIV}. This information is used in the *top-down architecture* to calculate the R and C values of the filter taking into account the device ratios $\frac{R_3}{R_2}$ and $\frac{C_3}{C_1}$ as well as the (possibly back-annotated) VCO input impedance defined by R_4 and C_4. The *bottom-up architecture* allows to directly specify the R and C values.

The *VCO model* constitutes to date the most complex model of the RF_TRX library. Five different interfaces (i.e., entities) provide the RF output signal at various levels of abstraction:

- **Instantaneous Frequency:** It is calculated using Eq. 10.5. This accelerates transient simulations as the integration step width of the analog solver can be increased up to one order of magnitude below the time constant of the PLL. This abstraction can be useful in combination with other base-band modeling approaches [2] to use the VCO output for up- and down-conversion in the frequency domain.

- **Single-Ended and Differential Digital Signals:** They are generated completely in the Discrete-Event (DE) domain by sampling $f_{\rm vco}$ at the state change of the digital signal and calculating from it the delay till the next state change. This decoupling of the analog and discrete solvers allows again the analog solver to increase its integration step width. This abstraction is useful if the VCO output is fed directly into digital blocks.

- **Sinusoidal Quantity or Differential Voltage Signals:** They are generated by integrating the instantaneous $f_{\rm vco}$ over time to calculate the phase, which defines together with the specified amplitude the current value of the sinusoidal waveform. This generation of the RF signal in the Continuous-Time (CT) domain slows down considerably the analog solver, which integration step-width needs to stay one order of magnitude below $f_{\rm vco}$. It is still more efficient than using a complete LC oscillator model. This precise abstraction may be necessary to interface analog blocks operating at RF and to do spectral analysis.

All five interfaces have in common that they use the instantaneous $f_{\rm vco}$ (Eq. 10.5). That is why it was decided to reuse the VCO model with frequency output inside the models with digital or sinusoidal output. The *frequency to digital* and *frequency to sinusoidal conversions* have been factorized into separate models, which can be reused independently.

The VCO model implements the functional specification from Section 2.0 on three abstraction levels. The *ideal* architecture implements the band switching capability and uses a constant frequency sensitivity $K_{\rm VCO}$. No input impedance is considered at the tuning terminals. Two *detailed* architectures (one for top-down design, the other for bottom-up verification) implement the full non-linear behavior of the VCO, as specified in Section 2.0, including band-switching, non-linear varicap, non-linear input impedance depending on $v_{\rm tune,eff}$, and calculation of the instantaneous $K_{\rm vco}$. The *top-down variant* uses a subset of the generics defined in the entity (Listing 10.1), which represent the component specification (e.g., number of bands, frequency range, $K_{\rm VCO,OPT}$ at $f_{\rm VCO,OPT}$) and the target technology (fitting parameters of Eq. 10.8) to calculate as secondary parameters in the architecture (Listing 10.2) the internal device parameters (e.g., $A_{\rm CV}$, $C_{\rm C}$, C_P, and parameters of the tank capacitance bank). The *bottom-up variant* uses directly the device parameters passed through the generics.

Listing 10.1 Entity declaration of the VCO with frequency output model

```
entity vco_freq_out is
  generic ( -- Primary parameters of each abstraction level
    F_OPT : real := 830.0e6; -- Center frequency to be optimized [Hz]
    K_OPT : real := 60.0e6; -- Frequency sensibility [Hz/V]
    F_MIN : real := 750.0e6; -- Minimum center frequency [Hz]
    F_MAX : real := 890.0e6; -- Maximum center frequency [Hz]
    N_BITS : positive := 3; -- Number of bits for band selection [1]
    N_BANDS : positive := 8; -- Number of bands
    RATIO_CC_CV0 : real := 4.0; -- Ratio of C_C and C_VO [1]
    RATIO_CP_CC : real := 0.01; -- Ratio of C_P and C_C [1]
    L_TANK : real := 5.0e-9; -- Tank inductance [H]
    R_IN : real := 20.0e3; -- Input resistance [Ohm]
    -- Technology constants of the varicap
    V_0 : real := 0.087; -- Offset voltage [V]
    D_CV0 : real := 6.2e-3; -- Varicap cap. density at V_0 [F/m**2]
    DDCV0_DV : real := -9.3e-3; -- Slope of D_CV at V_0 [F/(m**2 * V)]
    D_CV_AMP : real := 3.0e-3; -- Cap. density modulation amp. [F/m**2]
    -- Component parameters from real design for bottom-up verification
    S_CV_DESIGN : real := 3.63e-10; -- Varicap surface [m**2]
    C_TANK_MIN_DESIGN : real := 5.47e-12; -- Minimum tank capacitance [F]
    C_TANK_MAX_DESIGN : real := 8.08e-12; -- Maximum tank capacitance [F]
    DEBUG : boolean := false); -- Debug instance
  port ( -- Pin-accurate
    signal band : in std_logic_vector(N_BITS-1 downto 0)
                 := std_logic_vector(to_unsigned(0, N_BITS)); -- Band selector
    terminal power_p, power_m : electrical; -- Power supply terminals
    terminal tune_p, tune_m : electrical; -- Tuning terminals
    quantity f_out : out real); -- VCO frequency [Hz]
begin
  -- Interface-level assertions...
end entity vco_freq_out;
```

Several programmable divider models have been implemented [1] with a focus on a flexible interface (size of control word, digital input or instantaneous frequency input) and parameterizable digital behavior (e.g., minimal/maximal division ratio, control word treated as absolute division ratio or offset to minimal division ratio). For a two-stage divider architecture, a divider model with integrated control logic for a prescaler has been implemented.

For fractional division ratio control and noise shaping, a Σ-Δ modulator model describing a MASH structure [12] of first to third order on Register Transfer Level (RTL) (using **generate** statements) has been developed [1]. The number of bits representing the integer and fractional part can be adjusted making it easy to adjust the modulator resolution during architectural exploration to meet the frequency synthesizer specification and to reuse it in other projects.

The divider and Σ-Δ modulator models are primarily targeted to be used during the top-down design phase. During bottom-up verification, they can be replaced by the gate-level HDL models with detailed annotated delays exported by the synthesis and place & route tools.

Listing 10.2 Top-down detailed architecture of the VCO with frequency output model

```
architecture td_detailed of vco_freq_out is
  -- Definition of support functions...
  -- Calculation of secondary parameters...
  -- Quantity and Signals declaration...
begin
  -- Debug code and architecture-level assertions...

  -- Behavior implementation:
  -- Power supply
  i_power == 0.0;
  -- load at input
  c_in == 2.0 * c_v + C_P + C_C;
  v_tune_eff == (1.0 / c_in) * i_tune'integ;
  v_tune == R_IN * i_tune + v_tune_eff;
  -- Band switching
  c_tank <= C_TANK_MAX - real(to_integer(unsigned(band))) * C_TANK_STEP;
  break on c_tank;
  -- Varicap capacitance
  d_cv == D_CV_AMP * tanh((DDCV0_DV / D_CV_AMP) * (v_tune_eff - V_0)) + D_CV0;
  c_v == S_CV * d_cv;
  dcv_dvtuneeff == S_CV * (1.0 - ((d_cv - D_CV0) / D_CV_AMP)**2) * DDCV0_DV;
  c_vtot == 1.0 / (2.0 / C_C + 2.0 / (c_v + C_P));
  dcvtot_dvtuneeff == 0.5 * (C_C / (c_v + C_P + C_C))**2 * dcv_dvtuneeff;
  -- Output frequency
  f_out == 1.0 / (MATH_2_PI * sqrt(L_TANK * (c_tank + c_vtot)));

  -- Compute design data (e.g. performance estimates):
  -- Voltage sensitivity of VCO frequency
  k_vco == 2.0 * MATH_PI**2 * f_out**3 * L_TANK * (-dcvtot_dvtuneeff);
end architecture td_detailed;
```

The power consumption of these PLL components is not yet explicitly modeled as it is less important for the evaluation of the RF specification. Still the power supply terminals are included in each model, which test that the applied supply voltage has the correct polarity and remains in the specified range. This is important to facilitate the bottom-up verification of the connectivity.

Common to all models is the systematic usage of assertions to check for consistent parameterization of the model and compliance of the model state with the modeling assumptions. To facilitate code comprehension and maintenance, a consistent coding style is enforced through peer-review establishing a naming convention, consistent commenting of each declared constant/quantity/signal regarding its purpose and physical unit, encouraging the usage of common coding patterns for digital behavior, keeping the implementation of each model effect as local and independent from each other as possible, etc. All architectures are validated with the same test bench by using configurations, which implement automated measurements and checks for the various model behaviors. Simulation control scripts and waveform display configurations for each configuration ensure the reproducibility of simulation runs. Compilation and simulation of all models and test benches is handled through a common build system implemented for ADVance MS (ADMS). This reduces drastically the modification/simulation turn-around times easing the check of code modifications. The model library and its simulation environment are managed by a Subversion repository facilitating its parallel development and usage.

3. Usage of the RF_TRX Library in the Design Process of a Binary FSK Transmitter

The binary FSK transmitter used in this section as an example is based on the frequency synthesizer architecture depicted in Fig. 10.2 with direct modulation done at the Σ-Δ modulator input. The Σ-Δ control words are generated by a bit-stream modulator model based on the bit stream sampled at its input, to which it can apply a pulse-shaping technique (none, ramped, or raised-cosine) [1]. The PA is not modeled in this example, as the focus is on the evaluation of the modulation process.

Implementation of the FSK Transmitter Model

The transmitter model is realized as a structural model instantiating the necessary `RF_TRX` models not directly, but through **`component`** declarations provided in a **`package`**. All generic parameters of the component instances are forwarded through the **`generic`** declaration of the transmitter model. This allows us to parameterize the whole model hierarchy from the test bench. The constants used for this purpose are declared in an own package for each design case (target specification to implement). This has several advantages. The consistency during the evaluation of different architecture options is ensured, since constants, common to all architectures, are declared only once. Architecture-specific constants can be declared in the same package possibly based on the common ones. Constants representing the back-annotated component parameters from transistor level are declared in this package, too. The stimuli generation (in our case pseudo-random bit-stream, reset, clock, and power supply) is factorized into an own model and parameterized in the test bench by the design case constants package.

The test bench to be created for each design case and architecture to be evaluated gets thus very simple, as it just includes the constants package for the design case, declares all signals to interconnect the instances created of the stimuli generator and transmitter architecture models, and carries out the constants mapping to the generic parameters. Finally, the mapping of the components instances is done in a **`configuration`** for the test bench. This allows us, e.g., to reuse the same structural model during the top-down design and bottom-up verification phases by selecting the appropriate architectures of the component models. As for the component test benches, the transmitter model

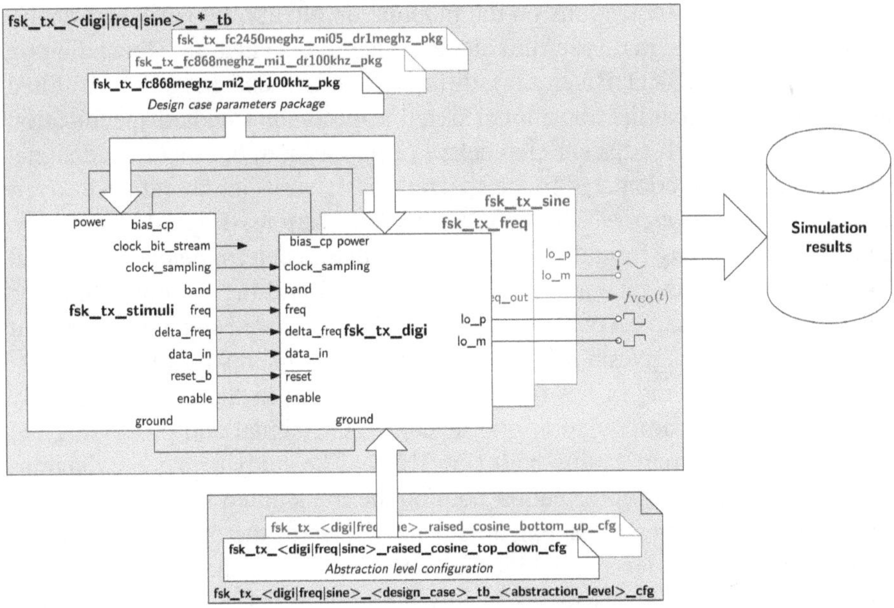

Figure 10.6. Organization of the FSK transmitter test benches

is integrated into the automated build system with simulation control scripts and waveform display configurations for each design case model configuration. This careful model organization (Fig. 10.6) achieves a full orthogonalization of the aspects *model structure*, *parameterization*, and *abstraction selection* avoiding code duplication as well as making it easy to add new design cases and model configurations.

Simulation of Different Transmitter Design Cases

Three different design cases have been chosen to validate the FSK transmitter model and show its flexibility. The first two examples represent possible wireless sensor network applications. The targeted output frequency (868 MHz) is in the Short Range Device (SRD) activity band and the chosen modulator data rate (100 kbit/s) allows to reach the duty cycle requirement (<1 %) for a low data rate application, e.g., a sensor data transfer. The two simulations differ in the modulation index, set respectively to 2 and 1, to analyze the influence of a frequency deviation modification on modulation. The third design case targets a transmission in the ISM band, for a compliant Bluetooth™ or Wibree™ application. The chosen carrier frequency and data rate are 2.45 GHz and 1 Mbit/s, respectively, suitable for applications like audio streaming. For this scenario, a modulation index of 0.5 has been used.

In order to have indications on the modulation quality, two different analysis are performed: signal spectrum and eye diagram evaluation. Depending on the targeted band (SRD, Bluetooth), different requirements have to be fulfilled in order to verify that the transmitted signal respects the standard specification and does not perturb adjacent channels. For this reason, the modulated signal spectrum has to be compared to the spectral mask, verifying that it never overcomes this reference curve. In the presented simulations (Fig. 10.7), only the contributions coming from the ideal PLL modulation determine the final signal spectrum. The noise generated by the PLL blocks and the PA non-linearities are thus not taken into account. Nevertheless the modulated signal at the output of the VCO already gives a fair indication of the occupied spectrum highlighting possible problems in the modulation process or in the PLL design. By applying an ideal demodulation on the same VCO signal and processing the result, the eye diagram can be built (Fig. 10.8). The study of the eye diagram allows to gather useful information both on the transmitted signal and the demodulation requirements. An example is the choice of the optimum sampling instant to minimize inter-symbol interference. Concerning the top-down design methodology, this analysis shows mainly the impact of the Σ-Δ quantization noise on the signal integrity, giving indications on the trade-offs between the reference, the carrier, and the filter cut-off frequencies. For the bottom-up verification phase, it is very convenient that the RF_TRX models can be also parameterized with the back-annotated device parameters – especially as the device parameters calculated by the top-down architectures cannot be implemented usually exactly on the transistor level due to design rules limitations and other constraints.

Table 10.1 summarizes the simulation performances obtained for the first and third design cases[1] of the FSK transmitter. Three transmitter model variants were developed, in order to investigate the impact of the different VCO output modeling approaches on the transmitter model simulation performance. As expected, the variant with digital VCO output performs better than those with frequency output (variant "frequential") and with sinusoidal output (variant "sinusoidal"), even though much more discrete events need to be processed by the digital simulator. The presented simulation results (Figs. 10.7 and 10.8) were obtained with the variant "digital". The "sinusoidal" variant is more precise, but much slower with respect to the "digital" variant (e.g., more than an hour of simulation time compared to ca. four minutes for case (a) in Table 10.1), while its spectra and eye diagram are very close to those of the "digital" variant. This sufficient precision and high simulation performance makes the RF_TRX library models very useful for the top-down architectural exploration and bottom-up verification phases. For comparison, a transistor

[1] The second design case is, in its specification and thus its resulting simulation performance, close to the first one.

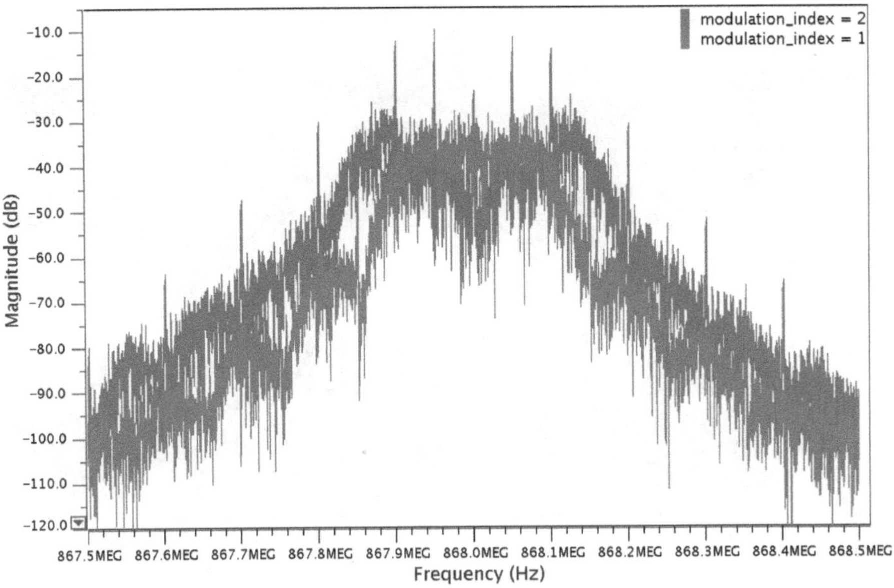

(*a*) $f_c = 868\,\text{MHz}$, $MI = \{2, 1\}$, $DR = 100\,\text{kbit/s}$

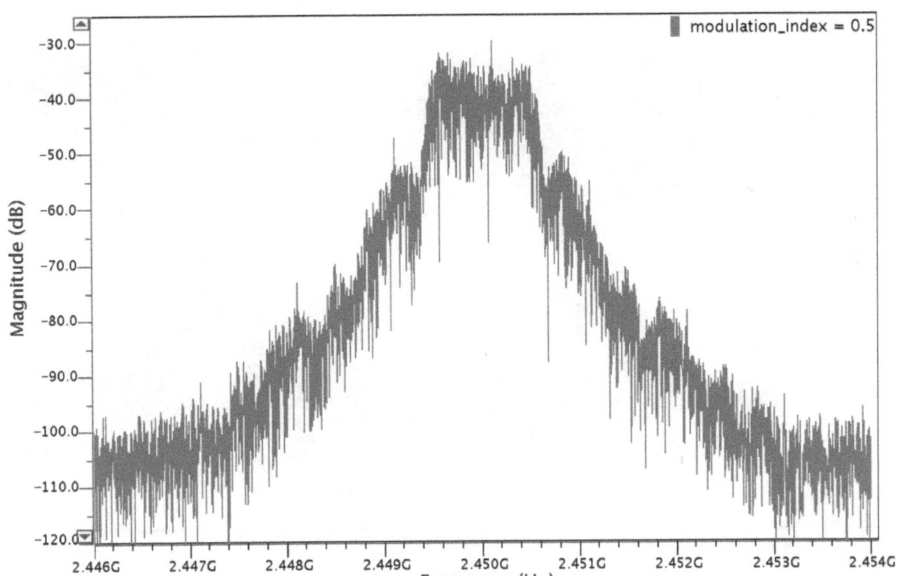

(*b*) $f_c = 2.45\,\text{GHz}$, $MI = 0.5$, $DR = 1MHz$

Figure 10.7. Frequency spectrum of the FSK transmitter output signal

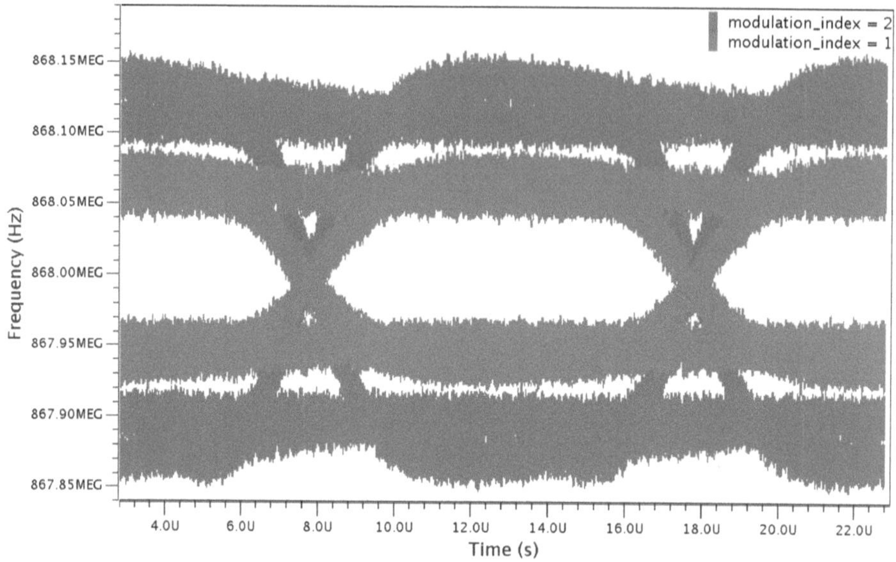

(*a*) $f_c = 868\,\text{MHz}$, $MI = \{2, 1\}$, $DR = 100\,\text{kbit/s}$

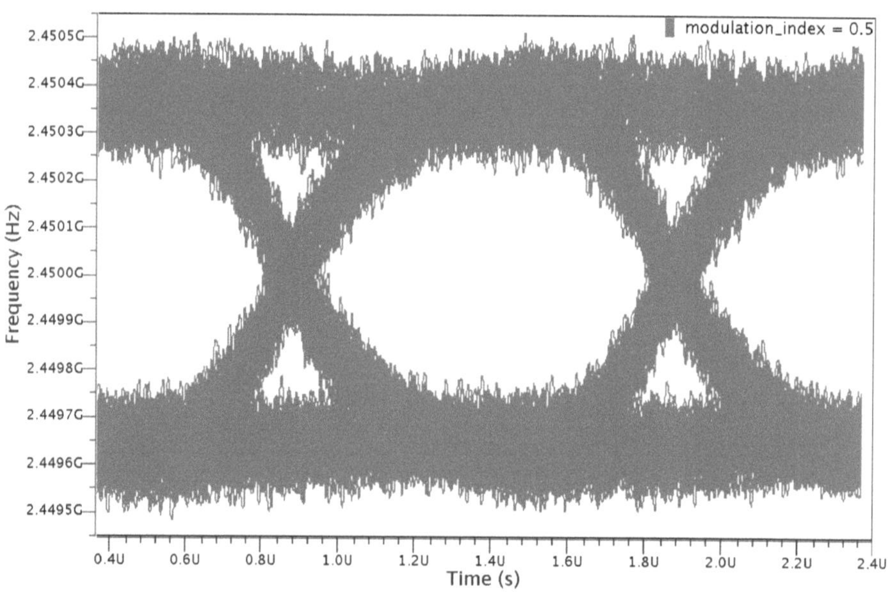

(*b*) $f_c = 2.45\,\text{GHz}$, $MI = 0.5$, $DR = 1\,\text{Mbit/s}$

Figure 10.8. Eye diagram of the FSK transmitter output frequency

Table 10.1. Simulation performance comparison of different FSK transmitter model variants run with ADMS 2008.2 [7] ($eps = 10^{-6}$) on an Intel Core 2 Quad 2.66 GHz CPU, 4 MB cache, 4 GB RAM, Linux2.6.9x86_64

10.1 (a) $f_c = 868$ MHz, $MI = 2$, $DR = 100$ kbit/s, $t_{stop} = 4$ ms

VCO output	h_{max}	CPU time	Steps	Events
Digital	1 µs	230 970 ms	2 637 297	44 173 736
Frequential	4 ns	281 240 ms	3 795 161	6 154 945
Sine	100 ps	4 598 380 ms	42 086 453	30 285 926

10.1 (b) $f_c = 2.45$ GHz, $MI = 0.5$, $DR = 1$ Mbit/s, $t_{stop} = 400$ µs

VCO output	h_{max}	CPU time	Steps	Events
Digital	100 ns	118 550 ms	911 153	14 935 101
Frequential	500 ps	144 410 ms	1 889 599	4 145 715
Sine	40 ps	1 232 120 ms	11 417 132	11 016 978

level simulation of the whole PLL (excluding Σ-Δ, modulator, and quartz oscillator blocks) has been tried. For a 10 µs transient simulation, the overall elapsed CPU time exceeded 24 h and after termination it was not possible to access the simulation results.

4. Conclusions and Outlook

This chapter has presented a methodology for modeling a complex RF system, namely a transceiver including a frequency synthesizer and the modulation for the transmitter part. In a first step, the behavior of the principal system components has been specified. Then, the corresponding models have been designed, implemented, and tested individually. Finally, the full system model has been assembled, thoroughly tested, and the simulation results presented. The development of the VHDL-AMS model library RF_TRX helped to establish best practices regarding communication between model developers and RF designers as well as organization and documentation of the models. Their application ensures a maximum flexibility, reusability, validity, and maintainability of the models. This approach shows that a complex RF system can be simulated rapidly and precisely by doing the right abstraction choices. The results show that such VHDL-AMS system simulations can provide the RF designer essential information, e.g., related to the spectral density or the eye diagram that transistor-level simulations cannot give in such a short time.

Future working directions include the consideration of different noise forms and the modeling of power consumption in the models as well as the development of new component models for the library to increase its coverage in the long term on the whole transceiver chain.

Acknowledgment

This work has been funded by the Hasler Stiftung under project № 2161.

References

[1] N.C. Bazungula, Development of a high-level VHDL-AMS simualation model of an RF transmitter based on a direct modulation architecture. Master's thesis, École Polytechnique Fédérale de Lausanne (EPFL), EPFL/STI/IEL/LSM, Station 11, CH-1015 Lausanne, Switzerland, 2009.

[2] J.E. Chen, Modeling RF systems. Technical report, The Designer's Guide Community. http://www.designers-guide.org/, visited on 3rd April 2009, 2005.

[3] R. Frevert, J. Haase, E. Jancke, *Modeling and Simulation for RF System Design* (Springer, Berlin, 2005).

[4] G.G.E. Gielen, CAD tools for embedded analogue circuits in mixed-signal integrated systems on chip. IEE Proc. Comput. Digit. Tech. **152**(3), 317–332 (2005).

[5] T.H. Lee, *The Design of CMOS Radio-Frequency Integrated Circuits* (Cambridge University Press, Cambridge, 1998).

[6] T. Mähne, A. Vachoux, ModelLib: a web-based platform for collecting behavioural models and supporting the design of AMS systems, in *Advances in Design and Specification Languages for Embedded Systems: Selected Contributions from FDL'06*, ChDL, chapter 4, ed. by S.A. Huss (Springer, Dordrecht, 2007), pp. 53–72.

[7] Mentor Graphics Corporation, *ADVance MS User's Manual: Release AMS 2008.2.*

[8] Mentor Graphics Corporation, *CommLib RF VHDL-AMS Library: Release AMS 2008.2.*

[9] N. Milet-Lewis, G. Monnerie, A. Fakhfakh, D. Geoffroy, Y. Herve, H. Levi, J.J. Charlot, A VHDL-AMS library of RF blocks models, in *Proceedings of the Fifth IEEE International Workshop on Behavioral Modeling and Simulation (BMAS) 2001* (Santa Rosa, CA, 2001), pp. 12–14, http://www.bmas-conf.org/2001/papers/bmas01-milet.pdf, visited on 7th June 2010.

[10] National Semiconductor, *An Analysis and Performance Evaluation of a Passive Filter Design Technique for Charge Pump PLL's.* National Semiconductor. application note 1001, http://www.national.com/an/AN/AN-1001.pdf, visited on 5th April 2009, 2001.

[11] J. Ravatin, J. Oudinot, S. Scotti, A. Le-Clercq, J. Lebrun, Full transceiver circuit simulation using VHDL-AMS. *Microwave Engineering*, 2002, pp. 29–33. http://www.bausch-gall.de/gsmtrans.pdf, visited on 30th April 2009.

[12] J. Rogers, C. Plett, F. Dai, *Integrated Circuit Design for High-Speed Frequency Synthesis* (Artech House, Boston, London, 2006).

Part IV

ASSERTION BASED DESIGN, VERIFICATION & DEBUG

The assertion of formal properties by means of standardized formalisms such as PSL and SystemVerilog Assertions provides a uniform expression of expected system behaviour. While traditional assertion-based design (ABD) is mainly focussed on the formal verification or co-simulation of assertions at the register-transfer level, the application of ABD has been extended considerably in last recent years. For instance, new methods were developed to employ ABD at the transaction-level and to automatically synthesize hardware from a set of assertions.

The following four contributions address advanced concepts for the support of assertion-based design and synthesis. A new method is shown for the synthesis of hardware from a set of "operation properties" which represent a transaction-level abstract view of an rtl-design. A re-use methodology for the verification of System-on-Chip protocols is demonstrated.

The construction and cosimulation of monitors from PSL-properties for SystemC transaction-level modeling is presented. Finally, a new method for improved constraint-based random stimuli generation for transaction-level models based on SMT techniques is introduced.

Hans Eveking
FDL'09 ABD Technical Area Chair
Technical University of Darmstadt, Germany

Chapter 11

HIGH LEVEL SYNTHESIS USING OPERATION PROPERTIES

Jan Langer and Ulrich Heinkel
Chemnitz University of Technology, Chair Circuit and System Design,
09126 Chemnitz, Germany
{jan.langer,ulrich.heinkel} @etit.tu-chemnitz.de

Abstract We propose a high level synthesis approach to generate RT level hardware from a specification of operation properties. The property language is called InTerval Language (ITL) and we assume the set of properties is complete, such that the properties alone are strong enough to map every possible sequence of input data to exactly one sequence of output data. A major advantage of using operation properties as a design method is the existence of commercial tools to check the completeness and consistency of the property set. Furthermore, operation properties are well suited for specifications of consecutive operations of finite length. We show the practicality of our method by implementing a particle filter for a localization application.

Keywords: Operation properties, high level synthesis, ITL, assertion-based design

1. Introduction

Over the last years, formal verification has become powerful enough to be applicable to large designs. Writing temporal properties to capture the behavior of a system is now common knowledge for both designers and verification engineers. Usually, properties are used as assertions in a traditional simulation-based verification strategy or they are formally proven by a model checker. Recently, they are also used to generate models that can monitor the correctness of a design during simulation or emulation. Alternatively, such models can represent an abstraction of specific aspects of the design to speed up simulation or to enable the simulation of components before they are actually implemented. A new idea is to derive the complete implementation of the design from a specification consisting of properties.

D. Borrione (ed.), *Advances in Design Methods from Modeling Languages for Embedded Systems and SoC's*, Lecture Notes in Electrical Engineering 63,
DOI 10.1007/978-90-481-9304-2_11,

A *Gap Free Verification* methodology based on operation properties has been introduced by the commercial tool 360MV™by OneSpin Solutions [9]. It provides a special property syntax known as InTerval Language (ITL). A set of additional rules helps to write a complete set of properties, that explicitly covers the design intent for every valid sequence of input values. The tool employs a powerful engine to prove the completeness of the property set as well as the correctness of each individual property with respect to the design. A property set is said to be complete if the conjunction of the properties alone is able to map every valid sequence of input data to exactly one corresponding sequence of output data [4, 6]. The completeness of a property set can be proven without the need of an actual design. It is possible to use PSL or SVA properties in a similar methodology. However, we concentrate on ITL, because, in contrast to PSL or SVA, it is specifically designed for specifying complete sets of properties.

In this work, we argue that using operation properties is a very good description method to design certain types of hardware components. The description of operation properties resembles timing diagrams in traditional specifications. Furthermore, we developed a synthesis tool that can generate very large circuits and is not prone to the state space explosion problem. We achieve this by avoiding the need to handle the circuit at the boolean level. We derive the control graph only from the structure of the properties and the data flow is directly mapped to hardware on RT level. Using the existing methodology and tools during the specification phase to prove consistency and completeness of the property set is a major advantage of using operation properties.

Writing a complete ITL specification based on operation properties is often a grey boxing approach, i.e. a certain level of insight into the inner workings of the desired design is required. This makes the specification less abstract than traditional Linear Time Logic (LTL) formulas. However, for the same reason it appears more appropriate to derive an actual implementation from ITL.

This paper is structured as follows: in Section 2 we introduce some of the previous work on synthesizing hardware from properties. In Section 3 we briefly show the structure and verification methodology behind operation properties. Section 4 describes the basic algorithm we use for our synthesis approach and in Section 5 we show two examples we have implemented. Section 5 summarizes this paper and gives a short outlook of future work.

2. Previous Work

Much effort has been put into the efficient generation of models of single properties used as monitors, or sets of properties describing the complete behavior of a component. Most of this work has been focused on a specification consisting of properties in LTL, mainly a subset of PSL [1–3, 5, 7, 10, 11].

All methods to synthesize LTL formulas are restricted in either the subset of supported constructs or the complexity they can handle. Another problem is the ambiguity of the specification. In most cases, a property or a set of properties is satisfied by more than one exact behavior. Thus, the synthesis method can either create a general solution that contains all consistent behavior or an arbitrarily chosen specific solution.

In [12] a specification consisting of properties in the form $A \rightarrow P$ are normalized and transformed into an executable design description.

3. Operation Properties

In order to understand our synthesis approach, it is necessary to understand the underlying verification methodology [8]. It uses a set of properties P, called operation properties. Each property $p \in P$ is a tuple $(A, C, t_{Left}, t_{Right})$ of assumptions A, commitments C and the two time points t_{Left} and t_{Right}, that indicate the start and end time of the property. An assumption is a function of all input values, output values and internal state variables of the component. It can use those values at every time point relative to the respective property and must evaluate to true or false. An assumption could be the function

$$g_{in}(t) + h_{in}(t+3) < 10, \tag{11.1}$$

that delays input g at time t, adds input h at time $t+3$ and checks if the result is less than 10. The time points are not absolute values. They always represent an offset to an arbitrarily chosen time t. The time point t_a at which an assumption a is completely evaluated is the maximum time point of all variable accesses in a, $t+3$ in the above example.

A commitment $c \in C$ is a function similar to an assumption, that evaluates not to a boolean value, but to a value that is assigned to an internal or output variable at a specific time point. An exemplary function that assigns the sum of two input variables to output o_{out} at time point $t+4$ is given by

$$o_{out}(t+4) \leftarrow g_{in}(t) + h_{in}(t+3). \tag{11.2}$$

Properties that are not intended to synthesize a design can use commitments of arbitrary form. However, our synthesis approach requires the above restrictions to generate valid design descriptions.

The time point t_{Last} of a property is the maximum time at which either an assumption is completely evaluated or a signal is assigned via a commitment. Furthermore, it must be greater or equal to t_{Right}.

A complete property describing the SYNC operation of a frame synchronization component (BITFRAMER) is listed in the algorithm of Fig. 11.1. It detects a synchronization word "1010" in a serial stream of input data and generates the correct output signals.

```
property sync is
  assume:
    at t : din = "1";
    at t+1 : din = "0";
    at t+2 : din = "1";
    at t+3 : din = "0";
  prove:
    during [t+5,t+16] : inframe = "1";
    during [t+5,t+16] : dout = prev(din);
    during [t+5,t+16] : frmpulse = "0";
  left_hook: t+3;
  right_hook: t+15;
end property;
```

Figure 11.1. SYNC property

This property will contain four assumptions at time points t to $t + 3$ and twelve commitments (at time points $t+5$ to $t+16$) for each of the three output signals. The time points t_{Left} and t_{Right} are $t+3$ and $t+15$ respectively. t_{Last} will be $t + 16$.

The described component has one input signal din representing the serial input data stream. Additionally there are three output signals. $dout$ is the serial output data stream, whereas $frmpulse$ signals the start of each completely synchronized frame. The $inframe$ signal is valid as long as the output stream is synchronized.

There may exist an operation $prev(d, n)$ representing the value of the variable d exactly n time steps before the time step specified with at or $during$. In contrast to the original verification methodology, we do not need to handle the internal state of the design in the properties. In case we need to prove the correctness of those properties against an implementation, we would have to add the necessary statements to our property before using the 360MV tool.

Additionally, a property graph is required that defines the successors of each property. For two consecutive properties, the time point t_{Right} of the predecessor is equivalent to the time point t_{Left} of the successor property. In case a property has more than one successor, the assumptions of the successor properties decide which property will define the design's behavior. A property graph for our BITFRAMER example is shown in Fig. 11.2.

After reset, the design always starts at the first time point of the reset property. Every possible sequence of input data must lead to one or more sequence of properties. However, in case there is more than one matching sequence, all sequences must result in the same sequence of output values. The difference

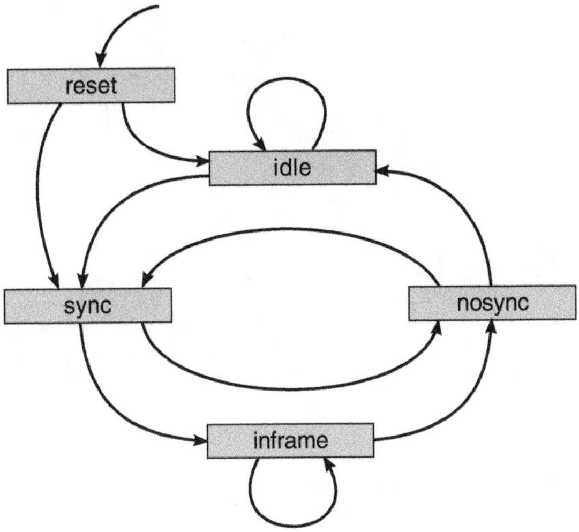

Figure 11.2. Property graph of the BITFRAMER component

to a traditional FSM is the fact that an operation property is usually active for more than one time point, and its duration differs from property to property.

4. Synthesis Algorithm

In this section we first propose an algorithm to synthesize the control flow of the design. After that we describe how assumptions and commitments form the corresponding data flow. During construction of the control graph, the commitments are not considered and the assumptions are simplified to boolean predicates. Furthermore, commitments are only allowed to assign variables at least one time point after the last variable access, in order to guarantee synchronous behavior, i.e. a commitment

$$z_{out}(t+2) \leftarrow x_{in}(t) + y_{in}(t+3) \qquad \text{(invalid)}$$

is prohibited because driving the output value of z at time point $t+2$ requires the value of y at time $t+3$.

The property graph of our BITFRAMER example is shown in a different representation in Fig. 11.3. Now, the time points of each property are unrolled. The question marks indicate the time points of the four assumptions employed in this example. For each property, the light grey and dark grey time points are t_{Left} and t_{Right}, respectively. The transitions between properties do not consume time and can be compared to ϵ transitions in conventional non-deterministic finite automata. However, there is a difference. The time steps after t_{Right} are executed, no matter which property follows.

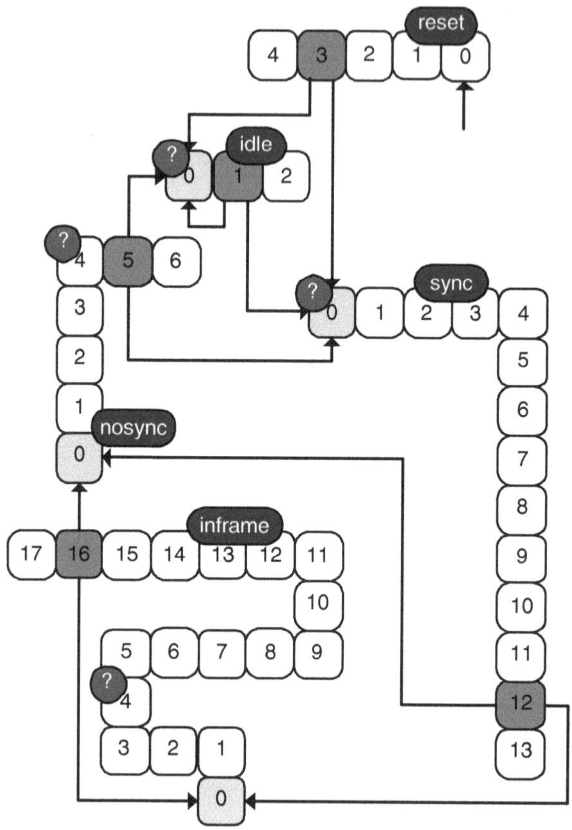

Figure 11.3. Unrolled property graph of the BITFRAMER component

In our example the overlapping between properties is relatively small. In case of a deep pipeline behavior of the design and short properties in relation to the pipeline depth, the overlapping can be significant and more than two properties can overlap. Thus, the property graph in Fig. 11.3 is not yet ready for synthesis. The following algorithm transforms it to a non-deterministic finite state machine that does not exhibit overlapping behavior. For the rest of the paper, a time point t of a property p is referred to as property position $p{:}\ t$.

The graph that is generated by this algorithm is defined as

$$H(S, E, A', C')$$

where S is the set of states of the new control flow graph and E is the set of transistions of this graph. All transitions are non-deterministic but no ϵ transitions are allowed. The relation $A' : S \mapsto 2^A$ annotates every state with a set of assumptions from the corresponding properties. The relation $C' : S \mapsto 2^C$ is similar for the commitments.

A state $s \in S$ is an ordered non-empty set of property positions. Every element $p\colon t \in s$ of this set defines that the state s has to exhibit the behavior described in property p at time point t. Especially all corresponding assumptions and commitments must be fulfilled. The order of the elements depends on the number of time steps since t_{Left} of the property. The *minimum* element of the state is defined as the property position with the shortest time passed since t_{Left}. The construction algorithm guarantees that exactly one property position is *active*, meaning that its time point is less than t_{Right}. Naturally, this element is also the minimum element in s.

The algorithm in Fig. 11.4 starts constructing the graph H by adding the initial state $\{reset\colon 0\}$ to a temporary set Q and iteratively searching for states following the already visited ones and adding the required transitions to E accordingly. New states that are found are stored in Q until they are visited. The algorithm terminates when no new states need to be added to S, i.e. the set Q runs empty.

In order to find the states following a given state s, the *successor(s)* algorithm in Fig. 11.5 is used. Basically, it increments the time points of every property position in s and drops the ones that exceed the corresponding t_{Last}. Afterwards, the minimum element is selected and in case it is the t_{Right} of its property, the state is split into different states for each successor of the property. The position t_{Left} of the corresponding property is added to each state.

In a next step the assumptions and commitments are annotated on the control graph. Figure 11.6 shows the result of our algorithm for the BITFRAMER

function $create$:
  $s_0 := \{\textbf{\textit{reset}}\colon 0\}$
  $S := \{s_0\}$
  $Q := \{s_0\}$
  while $Q \neq \emptyset$
    $s :=$ choose from Q
    $Q := Q \setminus s$
    forall $s' \in successor(s)$
      if $s' \notin S$
        $Q := Q \cup s'$
        $S := S \cup s'$
      end if
      $E := E \cup \{s, s'\}$
    end for
  end while

Figure 11.4. Function *create*

function $successor(s)$:
  $s' := \emptyset$
  forall $\{p\colon t_p\} \in s$
    if $t_p + 1 \leq t_{Last}(p)$
      $s' := s' \cup p\colon (t_p + 1)$
    end if
  end for
  $p\colon t_p :=$ min. element from s'
  if $t_{Right}(p) = t_p$
    $succ := \emptyset$
    forall $q \in P \mid \{p, q\} \in U$
      $s'' := s' \cup \{q\colon t_{Left}(q)\}$
      $succ := succ \cup s''$
    end for
    return $succ$
  else
    return $\{s'\}$
  end if

Figure 11.5. Function *successor(s)*

example. All transitions are non-deterministic and in general without conditions. The question marks on certain states indicate that the state has annotated assumptions. Commitments are not shown.

Usually, non-deterministic state machines are first transformed to deterministic ones and then mapped to hardware. However, to avoid the state space explosion problem of this transformation, we directly implement the non-deterministic graph in RT level hardware by means of the algorithm presented in [13, 14]. It encodes each state as a single bit variable. The variable is set to $'1'$ if any of its predecessor states is active and all of its annotated assumptions are fulfilled. The result is very efficiently generated but does not scale well in terms of flip-flop usage in case very long properties are needed.

The correctness of the result of the algorithm depends on the completeness checks by 360MV to guarantee a correct synthesis. This is because we implicitly assume those restrictions. As an example, the synthesized design will be wrong in case the assumptions of the IDLE and SYNC properties are not mutually exclusive.

The data flow components (assumptions and commitments) are mapped directly to RT hardware. ITL expressions and calls to macros are unrolled and flattened resulting in a plain netlist of ITL operators. We use a library of synthetic operators implemented in VHDL for all possible ITL constructs.

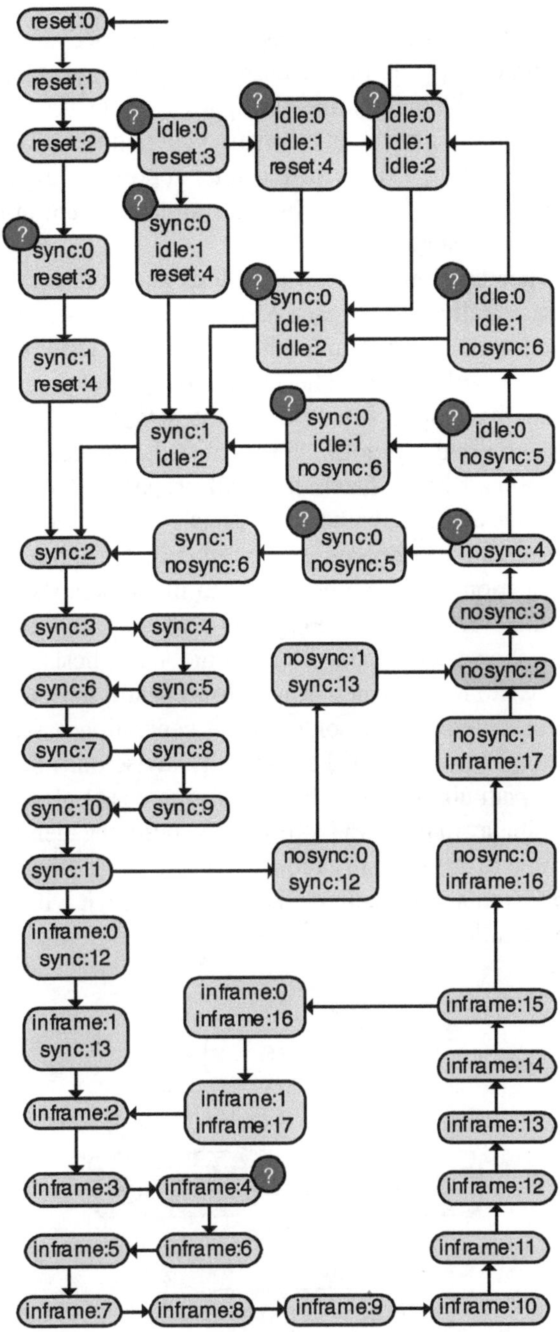

Figure 11.6. Generated control graph H of the BITFRAMER component

5. Results

In order to demonstrate the conciseness of the proposed design method, we specified and synthesized one of our current design projects using operation properties. The component implements a particle filter used for localization purposes [15]. A particle filter is an estimation technique for Bayesian models. Each of the 8,192 single particles of the filter represents a point in the state space that carries a weight. The weight is a simple estimate of the probability density function at that state point.

We split the specification into two partitions and connect them by FIFOs. The first step sequentially updates each particle and computes its weight. When all particles are updated and the sum of the weights is determined, a resampling step drops particles with low weight and clones particles with high weight. Figure 11.7 shows the general structure and the data flow of the particles.

The Updater component requires four properties to describe its behavior, whereas the Resampler block needs five. All properties have been written in just under three days and the synthesis to VHDL takes 1 second. The filter sequentially processes all 8,192 particles, a number that is appropriate for our 3D localization application. We have been able to implement the complete particle filter in a Xilinx Virtex-II Pro FPGA and it uses roughly half the available slices. Table 11.1 shows detailed information of the synthesis results. The maximum frequency is 4.3 MHz, which is quite low but still meets our design goals by a wide margin. One major problem of the generation has been the design of the weight computation. We had to carefully revise our initial properties to use efficient data path algorithms. The initial square root algorithm consumed about four times the available FPGA resources, but after identifying the problem, it was possible to dramatically reduce the resources. Once algorithms like square root or division are implemented as ITL macros, it will be easy to reuse them in other applications.

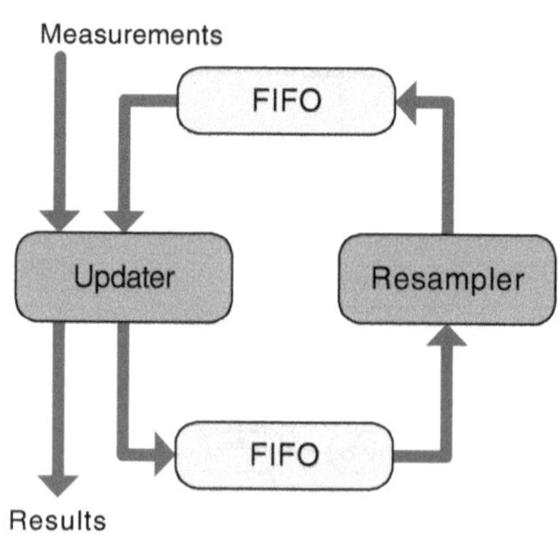

Figure 11.7. Structure of the particle filter

Table 11.1. Results of the synthesis algorithm for the particle design using operation properties on a Xilinx 2VP30ff896 device

Resource	Used	Utilization (%)
Flip-flops	2,835	9.76
CLB slices	7,741	56.52
Block RAMs	83	61.03
Block multipliers	113	83.09

The BITFRAMER component is a very simple abstraction of another motivating example for the property based design. The full framer component uses 16 Bit wide inputs and outputs and the frame length is about 20 cycles. The specification consists of five parameterized properties and the generation runs about half a second.

6. Conclusion

In general, the proposed method offers a very comfortable way to specify complicated timing diagrams for output signals. Often, the specification engineer thinks in operations and its very straightforward to implement those ideas in operation properties. Usually, it is simpler and less error prone to change the time points of signal assignments using properties than changing it in manually coded VHDL.

Future work will concentrate on finding and implementing a control flow synthesis algorithm that constructs a deterministic state machine instead of a non-deterministic one. This will open opportunities to implement an important ITL construct known as freeze variables more efficiently. This improvement will also enable our synthesis algorithm to handle very long properties, such as a synchronization framer of full functionality that motivated our initial effort for this tool.

The other weakness of our current tool is the method of generating VHDL. Some basic optimizations like sharing common logic or inserting RAM instances when necessary are not performed. For this step, we rely on the RTL synthesis tool, resulting in a high run time and memory usage.

The maximum frequency of circuits generated by our approach is often not yet satifactory. The implementation of the control flow in a deterministic state machine will enable further circuit optimization techniques that could increase performance.

References

[1] Y. Abarbanel, I. Beer, L. Gluhovsky, S. Keidar, Y. Wolfsthal, *FoCs – Automatic Generation of Simulation Checkers from Formal Specifications* (Springer Berlin/Heidelberg, 2000), pp. 538–542

[2] R. Bloem, S. Galler, B. Jobstmann, N. Piterman, A. Pnueli, M. Weiglhofer, Automatic hardware synthesis from specifications: a case study, in *Design, Automation and Test in Europe (DATE)*, 2007

[3] R. Bloem, S. Galler, B. Jobstmann, N. Piterman, A. Pnueli, M. Weiglhofer, Specify, compile, run: hardware from PSL. Electr. Notes Theoret. Comput. Sci. (ENTCS) **190**(4), 3–16 (2007)

[4] J. Bormann, H. Busch, Verfahren zur Bestimmung der Güte einer Menge von Eigenschaften, verwendbar zur Verifikation and zur Spezifikation von Schaltungen (Method for determining the quality of a set of properties, applicable for the verification and specification of circuits). European Patent Application issued 2005. Publication number EP1764715, 2005

[5] M. Boule, Z. Zilic, Efficient automata-based assertion-checker synthesis of SEREs for hardware emulation, in *Asia South Pacific Design Automation Conference (ASP-DAC)*, IEEE, 2007, pp. 324–329

[6] W. Buettner, M. Siegel, Achieving completeness in IP functional verification, 2007. http://www.eetimes.com/showArticle.-jhtml?articleID=197005268

[7] K. Morin-Allory, D. Borrione, Proven correct monitors from PSL specifications, In *Design, Automation, and Test in Europe (DATE)*, 2006, pp. 1246–1251

[8] M.D. Nguyen, M. Thalmaier, M. Wedler, J. Bormann, D. Stoffel, W. Kunz, Unbounded protocol compliance verification using interval property checking with invariants. IEEE Trans. Comput. Aided Design Integr. Circ. Syst. **27**(11), 2068–2082 (2008)

[9] OneSpin Solutions, 2009, http://www.onespin-solutions.com

[10] N. Piterman, A. Pnueli, Synthesis of reactive(1) designs, in *Verification, Model Checking, and Abstract Interpretation (VMCAI)*, 2006, pp. 364–380

[11] S. Ruah, D. Fisman, S. Ben-David, Automata construction for on-the-fly model checking PSL safety simple subset. Technical Report H-0234, IBM, 2005

[12] M. Schickel, V. Nimbler, M. Braun, H. Eveking, *An Efficient Synthesis Method for Property-Based Design in Formal Verification: On Consistency and Completeness of Property-Sets*, chapter 11 (Springer Netherlands, 2007), pp. 179–196

[13] R. Sidhu, V.K. Prasanna, Fast regular expression matching using FPGAs. In *IEEE Symposium on Field-Programmable Custom Computing Machines (FCCM)*, 2001

[14] R. Siegmund, *Ein Verfahren zur Spezifikation von Protokollen für die Verifikation und Synthese protokollorientierter digitaler Systeme*. PhD thesis, Technische Universität Chemnitz, 2005

[15] S. Thrun, W. Burgard, D. Fox, *The Particle Filter*, chapter 4.3 (MIT Press, 2005), pp. 96–113

Chapter 12

A RE-USE METHODOLOGY FOR FORMAL SOC PROTOCOL COMPLIANCE VERIFICATION

Minh D. Nguyen,[1] Max Thalmaier,[1] Markus Wedler,[1] Dominik Stoffel,[1] Wolfgang Kunz,[1] and Jörg Bormann[2]

[1] *Department of Electrical and Computer Engineering, University of Kaiserslautern, Kaiserslautern, Germany*
nguyen,thalmaier,wedler,stoffel,kunz@eit.uni-kl.de

[2] *Abstract RT Solutions GmbH, Munich, Germany*
joerg.d.bormann@web.de

Abstract We propose a new methodology for formally specifying on-chip bus protocols and for verifying protocol compliance of communication blocks in System-on-Chip (SoC) designs. In this methodology, the bus protocol is specified in a design-independent way by a set of protocol compliance properties based on a generic *recorder* finite state transition system. The properties are verified by combining local reachability analysis with a SAT-based property checking approach. This approach is called interval property checking and is based on a bounded circuit model generated from the design and the recorder.

The proposed methodology clearly differentiates between design-specific and protocol-specific aspects of the overall verification task and exploits the nature of typical SoC protocol specifications and implementations. In this way, the proposed methodology contributes to reaching two important goals: making the computational complexity of formal verification algorithms tractable for large designs and reducing the manual effort of applying formal methods in industrial practice. Our approach has been applied successfully on several industrial designs.

Keywords: Formal verification, protocol compliance, SoC, monitor, SAT, IPC

D. Borrione (ed.), *Advances in Design Methods from Modeling Languages for Embedded Systems and SoC's*, Lecture Notes in Electrical Engineering 63,
DOI 10.1007/978-90-481-9304-2_12,

1. Introduction

In recent years, formal property checking has become adopted successfully in industry and is used increasingly to verify individual modules in a System-on-Chip (SoC). In particular, Bounded Model Checking (BMC) [3] or related techniques have matured during the last decade and are well supported by industrial methodologies.

This paper is based on a SAT-based formulation of property checking called *Interval Property Checking (IPC)* [8, 9]. Similar to BMC, it is based on a bounded circuit model. The design and the property are unrolled into an iterative time frame expansion model to be handled by a SAT solver. However, in contrast to BMC no concrete initial state is specified. Therefore, global safety properties that formulate operations in intervals between abstract starting and ending states are proven unboundedly and very efficiently. It is important to note, however, that unlike BMC the checker may return false counterexamples corresponding to unreachable states since the bounded circuit model only captures local reachability and information about long-term dependencies may be missed. If this happens the properties need to be strengthened with reachability information, e.g., invariants, in order to rule out the spurious counterexamples.

IPC has been used successfully for formal property checking of *computational* SoC modules such as processors and hardware accelerators. However, it is sometimes less effective in handling *communication* modules which implement the interfaces between the computational modules. This is due to the fact that state machines of the communication modules have to handle transactions that may span over long time intervals and impose non-trivial reachability constraints on the state variables of such modules.

This paper proposes a methodology for verifying communication modules. It is based on IPC and primarily aims at reaching two goals: firstly, it reduces the algorithmic complexity of automatically determining sufficient reachability information to avoid false negatives. Secondly, it incorporates re-use concepts to further increase industrial productivity when verifying SoC interfaces.

Commercial providers of verification tools offer various concepts of design-independent verification IP for SoC protocols (IP = *intellectual property*). It should be noted, however, that all currently available solutions are intended for assertion-based verification flows. They only capture local behaviors and do not provide a complete coverage of the protocol. In our approach, transaction-style properties are formulated for the RTL implementation of the design. Our design-independent verification IP consists of a recorder and a set of IPC properties relating to it. Our approach is well adapted to the operational methodology of [8, 9] permitting a complete coverage of the design.

The recorder is a finite state transition structure (FST) used to remember previous bus states. It is connected with the design using only protocol-specific signals. The recorder FST is similar to a monitor as proposed in [11, 14]; it also watches the behavior of the bus signals. However, it does not check the correctness of the bus behavior and neither has a monitor output nor an error state. It is merely used as an auxiliary FST to formulate a set of compliance properties. It can be combined easily with IPC and is constructed in such a way that only little manual intervention is needed to compute invariants of sufficient strength without a full-blown reachability analysis. Based on this recorder, a set of generic compliance properties is formulated. Both, the recorder and the set of compliance properties are design-independent and need to be developed only once for a particular bus protocol. They can be re-used on any design implementing the same protocol.

As opposed to previous works, we do not use a monitor to check the compliance but use the recorder as a basis and guidance for formulating implementation-independent properties that completely describe the protocol-compliant behavior of the DUV. In contrast to the specification style proposed by Shimizu [11] which consists of a set of properties in an *implication style* of the form *past_condition* $\rightarrow$ *current_state*, our properties specify complete protocol transactions. Moreover, splitting the specification into a recorder and a property set allows for an intuitive way of describing the protocol behavior in terms of protocol states. Consequently, this style provides a natural partitioning of the verification problem that can be effectively exploited by the proof algorithms. It utilizes the strength of IPC in handling complex operational behaviors of communication modules. In addition, it also simplifies reachability analysis for IPC. In this way, we can prove complex properties including the data path of the communication modules.

Note that protocol compliance properties specifying the correct behavior of address and data signals in a complete protocol transaction cannot be handled effectively by other available formal property checking techniques such as k-step induction [10] and abstraction/refinement [6] even though they are less susceptible to the state explosion problem than conventional symbolic model checking. A conceptual analysis of the above approaches quickly reveals their limitations when being applied to properties that prove protocol-compliant behavior within a certain protocol transaction. For example, abstraction/refinement approaches should consider an abstract model of the DUV that consists of at least the signals appearing in the property – this comprises not only control but also address and data signals. Unfortunately, this abstract model is often beyond the ability of the model checking techniques used in current abstraction/refinement approaches. Moreover, in our observation, this abstract model still needs to be refined to prove the property, thus, further increasing the computational complexity. In k-step induction, an iterative circuit model consisting of $k+1$ time frames of the design is analyzed using SAT. The induction depth k, however, will become very large, in cases where subtle dependencies between the state variables of a design exist and where the property does not reflect these dependencies. Unfortunately, communication modules typically perform a lot of bookkeeping in order to correctly perform the (long) transactions of a protocol. This results in such dependencies and prevents k-step induction from effectively proving the property. We expect that also property checkers based on k-step induction will benefit from the here proposed techniques for reachability analysis.

The rest of this paper is organized as follows. In Section 2, the overall verification methodology is presented. In Section 3, we define the recorder and describe a systematic procedure for designing it as well as for specifying the bus protocol. In Section 4, the proposed approach is experimentally evaluated. Finally, a conclusion is presented in Section 5.

2. Verification Methodology

In this section we present our proposed verification flow for protocol compliance, as depicted in Fig. 12.1. The flow consists of two parts. The first part, shown on the right-hand side, produces a design-independent verification IP and will be presented in detail in Section 3. The resulting IP consists of a recorder and a property suite. The second part of the flow is design-specific and consists of two major steps. First, the recorder is connected with

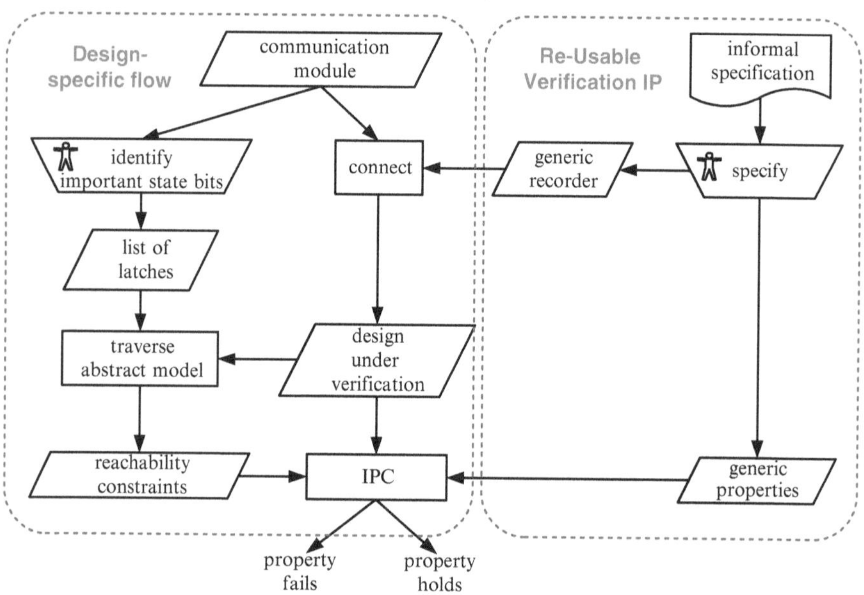

Figure 12.1. Protocol compliance methodology

the implementation to construct the design under verification (DUV). Second, reachability constraints are generated in a semi-automatic process as depicted in the left part of the flow in Fig. 12.1. In this flow the user guides the tool in the abstraction process by specifying a list of design latches that he considers to be relevant for the protocol behavior. The generated reachability information is used to rule out false negatives that otherwise may occur when the generic protocol compliance properties are checked using IPC [8].

For proving interval properties on the resulting DUV we use a SAT-based interval property checker (IPC) [8]. The property checker produces unbounded proofs for interval properties. The presentation of this basic technique for property checking is beyond the scope of this paper. The interested reader is referred to [8]. An important characteristic of IPC is the fact that the checker may return spurious counterexamples that start in an unreachable state of the design. In order to rule out such false negatives the generic properties may be augmented with additional reachability information.

To keep the manual verification effort small it is desirable to generate this information automatically. Unfortunately, for complexity reasons, an exact reachability analysis is often impossible for the type and size of designs considered here. Traditional symbolic model checking techniques cannot be used. Therefore, similar as in [5], we use approximate reachability information to restrict the state space. In our case, however, after introducing the recorder the reachability information needed for IPC can be obtained more easily. In [8], reachability information is generated by traversing a large number of small abstract models of the DUV. We use this approach in the design-specific part of the verification flow of Fig. 12.1. As it turns out, in our case, traversing a single abstract model is sufficient. This abstract model is derived from the DUV by localization abstraction (e.g., [7]): all latches are abstracted except for the ones belonging to the recorder and those latches of the design that directly relate to the bus control signals. The abstraction typically does not contain any data signals.

Since our abstract model contains only latches related to the bus control signals the verification engineer needs to inspect only a small portion of the design to identify these latches. Typically, the verification engineer needs to consider design latches that are (i) in the cone of influence of the bus control signals; (ii) the important control state variables of the design. This manual step is affordable in practice. Nevertheless, an automated solution is desirable. We are currently investigating ways to automatically identify the design latches constituting the abstract model [12].

Finally, note that in contrast to abstraction refinement techniques (e.g., [6, 13]) we do not check properties on the abstract model; instead, the abstraction is only used to generate sufficient reachability information for IPC proofs.

3. Formal Protocol Specification

In bus protocol specification documents, e.g., the AMBA specification [2], the concept of a *bus state* is often used. A bus state represents the history of the bus signals. It contains sufficient information to determine all possible current behaviors. Protocol-compliant bus behavior can often be described in a form like *"if the bus state is IDLE, a data or an IDLE transaction can be started with HTRANS=2 or HTRANS=0, respectively"*. Bus states are therefore essential to describe the bus behavior. They are usually identified and specified before properties for the bus are written. However, it is hard to specify bus states using only the current bus signals because a bus state represents information about the history of the bus. For example, in order to specify that the bus is in a specific bus state *IDLE*, all possible prior sequences of the signal values on the bus leading the bus to state *IDLE* need to be specified. We suggest to use a recorder to store the history of the control signals of the bus in so called *recorder states*.

We partition the formal specification of the bus into two parts: (i) the recorder, representing the control behavior of the bus; (ii) a set of properties, representing the behavior of the remaining signals such as the data lines. For simplification, we use the term *bus behavior* for both control-related and data-related behavior in the sequel.

In the remainder of this section, we will first give a precise formal definition of the recorder and illustrate the concept by means of a concrete example. We then discuss a systematic (manual) synthesis procedure for such a recorder. This is a creative step because it formalizes parts of the informal protocol specification. We will take an intuitive, top-down approach that provides us first with a bus state-related "skeleton" in the form of a non-deterministic FST that is then extended using auxiliary FSTS to yield the final deterministic recorder.

Bus Recorder

Consider a bus with a set of control signals $\hat{X} = \{\hat{x}_1, \ldots, \hat{x}_n\}$. These control signals have a finite Boolean or natural domain and encode an alphabet $\Sigma = \{\sigma_0, \ldots, \sigma_m\}$ of m *bus symbols*. The control-related bus behavior can be considered as a *bus language* $\mathcal{L} \subset \Sigma^*$. The recorder, which is related to the bus language, is defined as follows:

DEFINITION 12.1 *Given a bus with a set of bus symbols Σ, a recorder for a bus is a labeled, incompletely specified, finite state transition structure (FST) $(\Sigma, S, s_0, \delta, L)$, where*

1. *Σ is the set of input symbols (bus symbols).*
2. *S is the set of* recorder states.
3. *s_0 is the initial state.*

4 $\delta : S \times \Sigma \mapsto S$ *is the incompletely specified transition function.*

5 $L : S \mapsto 2^{\Sigma}$ *is a labeling function.*

such that for any possible word $w = \sigma_0\sigma_1 \cdots \sigma_t \in \mathcal{L}$*, there exists a sequence of recorder states* $(s_0, \ldots, s_t) \in S^{t+1}$ *such that* $\delta(s_i, \sigma_i) = s_{i+1}$ *for* $0 \leq i < t$*. The labeling function assigns to each state* $s \in S$ *the input symbols* $L(s)$ *that are allowed to occur on the bus (according to the protocol specification) when the recorder is in state* s*. The transition function* $\delta(s, \sigma)$ *is defined for every state* s *and for every input symbol* $\sigma \in L(s)$ *possible in that state* s*, and undefined otherwise.*

We illustrate Definition 12.1 using the AHB protocol as an example. In this protocol, there are some control signals *HTRANS*, *HGRANT*, *HRESP* and *HBURST* that encode the alphabet Σ of bus symbols. In the example, we only look at a small fraction of the recorder concerned with the *IDLE* state and the *INCR4* burst transaction, where 4 consecutive data packets are transferred.

Figure 12.2(a) shows a fragment of the state transition diagram of a recorder for the AHB protocol. This fragment focuses on the *INCR4* burst transaction. In the figure, the recorder states are labeled with the set of symbols $L(s_i)$. For example, when the recorder is in state s_0 (the bus is idle), there are two possible transactions: Either a data transaction can be started with *HTRANS*=2, or the *IDLE* transaction is continued with *HTRANS*=0. In both cases, the slave is expected to respond with *HRESP*=0. Hence, the set of symbols that may appear on the bus when the recorder is in its initial state s_0 is represented by the characteristic function $\Sigma_0 := (\mathit{htrans} = 0 \vee \mathit{htrans} = 2) \wedge (\mathit{hresp} = 0)$.

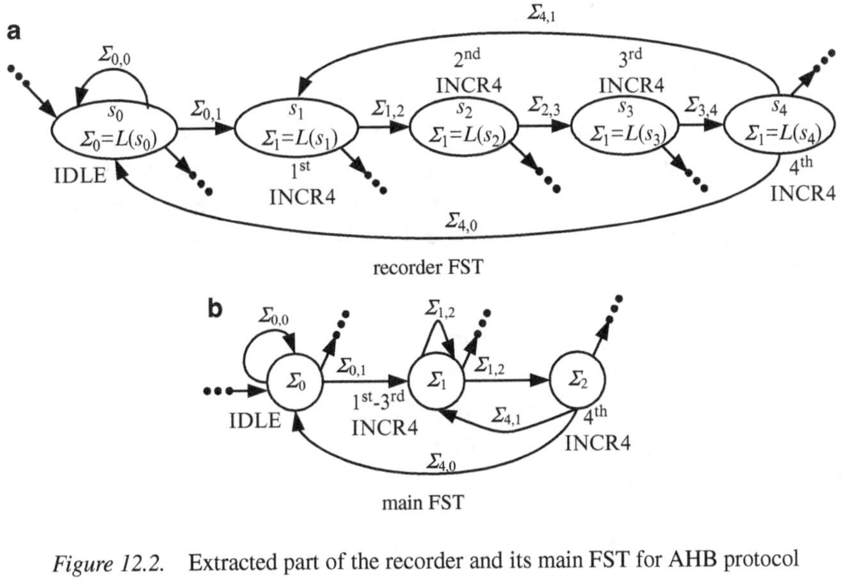

Figure 12.2. Extracted part of the recorder and its main FST for AHB protocol

In the same way the bus symbols expected in the four phases of an *INCR4* burst transaction are encoded by $\Sigma_1, \ldots, \Sigma_4$, where $\Sigma_1 = \Sigma_2 = \Sigma_3 := (\mathit{htrans} = 3 \vee \mathit{htrans} = 1) \wedge (\mathit{hburst} = 3)$, and $\Sigma_4 := (\mathit{htrans} = 0 \vee \mathit{htrans} = 2)$. In addition, the state transitions are marked with triggering conditions $\Sigma_{i,j}$. For example, the recorder takes the transition (s_0, s_1) if the *INCR4* transaction is started. This is the case if the condition $\Sigma_{0,1} := (\mathit{htrans} = 2) \wedge \mathit{hgrant} \wedge (\mathit{hresp} = 0) \wedge (\mathit{hburst} = 3)$ is fulfilled.

A recorder as defined in Definition 12.1 must be able to recognize all possible words of bus symbols that may occur on the bus. It should be constructed such that its states can be used to represent bus states. In this case a property can easily specify a bus transaction based on a sequence of recorder transitions between specific bus states. In contrast to a conventional monitor we do not require that forbidden words (words not in $\mathcal{L}$) are handled by the recorder. It is not the task of the recorder to recognize forbidden words. The recorder only needs to make defined transitions for correct words. It is left to the generic property set to ensure that only allowed symbols appear on the bus. This will be discussed in Section "Properties".

Based on Definition 12.1 we define a systematic procedure for recorder specification consisting of two phases. In the first phase, we focus only on the individual "situations" that the bus can be in, i.e., we identify sets of bus symbols that will be used as labels in our recorder later. Simultaneously, we determine the possible transitions between the labels. This leads to a non-deterministic FST which controls the overall behavior of the recorder. We call it the *main FST* of the recorder. In a second step, we add detail in the form of additional sub-FSTS communicating with the main FST to remove the non-determinism. This step yields the recorder. The recorder constructed in this way fulfills Definition 12.1 and represents the bus states in a design-independent way.

We suggest using the systematic (manual) procedure in Fig. 12.3 to determine the main-FST of the recorder. The concept of a main FST is best explained using an example. Figure 12.2(b) shows a fraction of the labeled state transition graph of the (non-deterministic) main FST that will be used to construct the recorder for the AHB protocol as illustrated in Fig. 12.2(a). As we formalize a more or less informal specification if protocol compliance, such a main FST can only be derived manually from the protocol specification document. We suggest using the systematic procedure in Fig. 12.3 for this purpose. Beginning with a specific starting configuration Σ_0 of the bus (for example, the situation where the bus is idle), the procedure systematically explores new sets of bus symbols Σ_i that describe a certain, important, control "situation" on the bus that can be reached in the next transition. The identified bus symbols Σ_i constitute a new state of the main FST. Finally there is a one-to-one mapping between the labels Σ_i and the states of the main FST of the recorder.

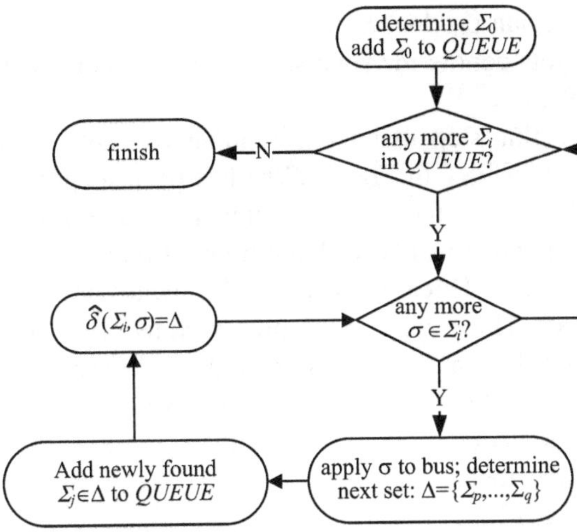

Figure 12.3. Constructing the main FST of the recorder

The starting configuration Σ_0 forms the initial state of the main FST and is entered into a queue. We determine the recorder states in a breadth-first manner. In each step of the search the procedure picks the first state Σ_i in the queue. For this main state/label Σ_i, we determine all possible next-cycle labels that can be reached from any bus symbol $\sigma \in \Sigma_i$. (This is done by evaluating the protocol specification document.) Because a set of bus symbols Σ_i by itself is not sufficient to capture the state of a transaction on the bus, for any given σ there may be several labels, i.e. a set Δ of labels, reachable in the next cycle. In other words, the transition function $\hat{\delta}$ of the main FST is that of a *non-deterministic* function. After we have identified the set of next-state labels $\Delta = \{\Sigma_p, \ldots, \Sigma_q\}$ for a given state Σ_i and bus symbol σ, we enter all the new labels $\Sigma_j \in \Delta$ into the queue, as every new label forms a new state of the main FST. Also, new transitions are added by updating the transition function of the main FST with $\hat{\delta}(\Sigma_i, \sigma) := \Delta$. For our running example the above procedure yields the state transition graph of the main recorder FST as illustrated in Fig. 12.2(b). Note the non-determinism in main state Σ_1 as there are two transitions to next-state labels, Σ_1 and Σ_4, that are triggered by identical sets of bus symbols.

The non-determinism originates from the insufficient information given by the set of bus symbols alone. In the second phase of the extraction procedure, we now extend the main FST by adding auxiliary sub-FSTS to the system such that the non-determinism is removed. The state set of the recorder is encoded in the state variables of the main FST and its sub-FSTS respectively, i.e., the state set S is the cartesian product $S \subseteq \hat{S} \times \tilde{S}_1 \times \tilde{S}_2 \times \cdots$ of the state sets $\hat{S}$ and $\tilde{S}_i$. These state sets denote the state sets of the main and sub-FSTS of the recorder, respectively.

We also illustrate this step using our running example of the AHB bus recorder. After the main FST in Fig. 12.2(b) has been identified, we need to add detail in order to remove the non-determinism in main state Σ_1. The STG in Fig. 12.2(b) corresponds to an *INCR4* burst transaction which consists of four consecutive data transfers. Three of these transfer cycles are identical with respect to the bus signals. These cycles are represented by Σ_1, which is a label for three consecutive recorder states. To implement these three states we add a sub-FST implementing a counter *cnt* with the state set $\tilde{S} = \{1, 2, 3\}$. The state set of the recorder now includes three distinct states $s_1 = (\Sigma_1, 1)$, $s_2 = (\Sigma_1, 2)$ and $s_3 = (\Sigma_1, 3)$, all labeled with the bus symbol set Σ_1. This leads to the recorder shown in Fig. 12.2(a).

Properties

In the second step of our methodology we specify the intended, compliant behavior of a communication module with respect to the considered protocol. This behavior is formalized by means of properties that reason about the recorder and the bus-signals. In principle this can be conducted using any property specification language. With the resulting properties, the design can be checked for correctness of its input/output behavior with respect to the bus, i.e., when referring to the inputs and outputs of the design in our properties we directly refer to the bus signals. We also refer to the states of the recorder as this circuit keeps track of the history of the bus signals. By this concept, the properties can be easily formulated without explicitly referring to internal design signals. This is important for portability of the verification IP to new designs.

In the following, we use $V = (\hat{V}, \tilde{V})$ to denote all state variables of the recorder. The states of the main FST are encoded by $\hat{V}$ and the sub-FSTS are encoded by $\tilde{V}$, respectively. In addition, the control and data bus signals are denoted by $\hat{X}$ and $\bar{X}$; the set of bus signals is: $X = \hat{X} \cup \bar{X}$.

The properties describe (i) the control-related and data-related behavior of the bus signals, and (ii) the state sequences of the recorder. The general property format used is that of a safety property of the form $\mathsf{AG}\,(a \rightarrow c)$. In the implication, a is called *assumption* and c is called *commitment*. Both, the assumption a and the commitment c, are so called *timed boolean expressions*, i.e., Boolean expressions over design signals that are potentially delayed using the temporal next state operator X. Properties of the above mentioned type specify a design operation of bounded length which can be associated with a series of transitions taking the system from one important bus state to another.

In our property set every property describes a certain transaction performed on the bus. Since the history of the bus is represented by the recorder states we use the state variables of the recorder to specify a *starting assumption* and an *ending commitment* as conditions for the bus signals at the beginning and at the end of the transaction under consideration. When specifying recorder states, we can explicitly refer to the state variables of the main FST of the recorder and to those of its sub-FSTS.

In practice, properties are written in the property language supported by the verification tool. In the following, we use the proprietary language ITL (InTerval Language) [9] of the IPC property checker we used in our experiments. Figure 12.4 shows a property specifying a bus transaction corresponding to a sequence of recorder main states $(\hat{s}_1, \ldots, \hat{s}_{n+1})$.

Line 2 specifies the starting assumptions for the main-FST, $(\hat{V} = \hat{s}_1)$, and the starting assumption, $a(\tilde{V})$, regarding the states of the sub-FST of the recorder. The set of bus symbols that trigger the main transition $\hat{s}_1 \rightarrow \hat{s}_2$ and initiate the bus transaction to be checked by the property is represented in line 3. The bus symbols triggering intermediate state transitions on the path $(\hat{s}_2, \ldots, \hat{s}_{n+1})$ are partially constrained by environment conditions $a_{j,j+1}(\hat{X})$, where $2 \leq j \leq n$ (lines 4–6). These conditions are determined by the behavior of the other bus participants but not by the design under verification. (Therefore, they are also not a part of the commitment to be checked.)

property protocol_behavior;
1: **assume:**
2: **at** t: $(\hat{V} = \hat{s}_1) \wedge a(\tilde{V})$; // starting assumptions
3: **at** t: $\Sigma_{1,2}$; // trigger condition
4: **at** t+1: $a_{2,3}(\hat{X})$; // environment conditions
5: ...
6: **at** t+n-1: $a_{n,n+1}(\hat{X})$; // environment conditions
7: **prove:**
8: **at** t+1: $c_2(\bar{X})$; // data commitments
9: ...
10: **at** t+n-1: $c_n(\bar{X})$; // data commitments
11: **at** t+1: Σ_2 ; // control commitments
12: ...
13: **at** t+n-1: Σ_n ; // control commitments
14: **at** t+n: $(\hat{V} = \hat{s}_{n+1}) \wedge c_e(\tilde{V})$; // ending commitment
end property;

Figure 12.4. Protocol compliance property

Lines 8–10 specify protocol-compliant behavior of the data signals, $c_j(\bar{X})$, where $2 \leq j \leq n$. Remember that the recorder was specified only for protocol-compliant bus behavior; it does not check correctness of the bus control signals – this has to be verified in the properties. Hence, the constraints in lines 11–13 check that when the recorder is in state $\hat{s}_j$, the bus control signals indeed take on only values from Σ_j ($2 \leq j \leq n$), ensuring that the recorder is only "fed" with allowed bus symbols when reaching state $\hat{s}_j$. Finally, line 14 specifies the ending commitment which states that the recorder has reached its ending main state $\hat{s}_{n+1}$ and, also, the sub-FSTS of the recorder fulfill their ending conditions $c_e(\tilde{V})$.

At this point, we have presented the method for specifying bus behavior using the recorder and a set of properties. By splitting the specification process into these two steps and by defining the recorder formally, we can construct the formal specification model in a systematic way. This helps to reduce the manual effort of formal specification, which is essential in any verification methodology. Also it should be noted that the recorder does not drive any input signals of the DUV, i.e., the DUV is not constrained by the recorder. This also differentiates our specification model from previous specification methods.

4. Experimental Results

We applied the proposed methodology for verifying three industrial and one public-domain [1] AHB master designs. All experiments were performed on a Core2Dual-PC with 2.13 GHz clock frequency and 8 GB RAM running openSUSE Linux 10.3.

Recorder Versus Monitor

We have constructed a recorder to verify protocol compliance of masters in the AHB protocol as described in Section 3 using the hardware description language *VHDL*. We have pointed out that it is beneficial to design a recorder that only reads the bus control signals as opposed to designing a monitor which also checks the correctness of transactions and, thus, also relates to the data path of the communication module. By separating the generic protocol specification in a recorder and in a set of properties we preserve the natural distinction between control and data path. In order to measure the effects of this we have extended the proposed recorder design to a full monitor, i.e., an error state was added to the recorder FST such that the wrong control- and data-related behaviors, i.e. the forbidden words, force the recorder to go to the error state.

We report the size of the recorder and the monitor in Table 12.1. As expected the monitor is considerably larger. Moreover, it relies on more bus signals. As a result, the SAT instances generated by the interval property checker grow by a factor of up to 3x.

Table 12.1. Size of recorder and monitor

	Recorder	Monitor
Lines of code	583	666
Inputs	11	47
Latches	15	51
Gates	3,647	13,923

Both the recorder and the monitor are each connected with a design to form a DUV. Since we use a recorder and a monitor eight different designs under verification are considered.

Checking Protocol Compliance with Standard Property Checking Techniques

For illustrating the relevance of the proposed techniques we applied standard property checking techniques in order to prove protocol compliance of the designs in our benchmark set. For this purpose we used the monitor-based properties described above. For the composed DUV generated from design and monitor we used VIS [4] to prove that the monitor never goes to the error state, i.e., $\mathsf{AG}\,(\hat{V} \neq \mathit{error})$. For the experiment we used every available exact and approximate approach for property checking implemented in VIS.

The experiment turned out that none of the approaches solved any of the verification problems. They were either interrupted due to time-out (>5 h) or memory-out (>2 GB), or they reported a false-negative result.

We also tried to prove the properties by k-step induction. However, we were not able to prove the property because of false-negatives even though we used an induction length of 100.

The results of this experiment confirm our conceptual analysis in Section 1 in that abstraction/refinement techniques and k-step induction may fail when proving protocol compliance properties for large designs.

Checking Protocol Compliance Using IPC and Reachability Analysis

We report results on checking the protocol compliance of the AHB designs with our generic verification IP. We provide data on both steps of the verification algorithm, namely the invariant generation and the IPC proof.

Reachability Analysis. For invariant generation we use the "TBT traversal" algorithm of [8]. We apply this technique to localization abstractions of the DUV with respect to the recorder state variables and manually selected important control state bits of the design.

Table 12.2. Experimental results

Design under verification		Size		TBT		IPC	
		# latches	# gates	CPU	Mem.	CPU	Mem.
fcdp-ahb	R	661/76	20,366	2,069	294	27	149
fcdp-ahb	M	697/76	29,157	6,557	301	36	188
ahb-master	R	3,388/36	109,421	70	136	293	530
ahb-master	M	3,429/36	119,674	102	138	6,109	6,897
bvci2ahb	R	161/55	6,838	300	138	256	480
bvci2ahb	M	197/55	17,239	410	119	609	963
open-ahb	R	466/32	15,656	578	488	194	489
open-ahb	M	502/32	25,896	462	390	130	169

Since the resulting abstract models are fairly small any conventional reachability analysis algorithm can be used as an alternative to TBT traversal as well.

However, for TBT is is sufficient to provide the selected design registers as sub-FSM for the recorder and the entire abstraction process will be done implicitly. The algorithm then traverses the sub-FSM together with the main FST of the recorder and identifies reachability constraints. The resulting constraints are added as an invariant to the IPC SAT instances.

The results are shown in Table 12.2 which is organized as follows. For each design we report the results for two DUVs: one DUV consisting of the design and the recorder (marked "R") and another DUV consisting of the design and a monitor (marked "M"). We report the DUV sizes in terms of the number of latches and the number of gates. In the column labeled "latches", we also report the numbers of selected latches in the abstract models. In the columns labeled "TBT" we report the CPU time (in seconds) and memory consumption (in MB) of TBT traversal when it traverses the abstract model.

Verifying Properties Using IPC. After the reachability analysis, we use the commercial IPC tool [9], i.e., Onespin360 MV™, to verify a set of 28 properties that completely covers the AHB protocol. The columns labeled "IPC" in Table 12.2 show the total CPU time (in seconds) and the peak memory usages (in MB) needed by the IPC tool to prove the set of properties.

As shown in Table 12.2 the algorithms spent less time and memory for the recorder with properties than for the monitor approach in three of the four designs. The results for the fourth design are due to a confirmed bug in the design. While the monitor immediately reveals this bug several of the recorder properties are proven to hold before the first failing property is processed and the bug is exposed.

5. Conclusion

This chapter contributes a practical methodology to verify SoC communication modules. It is an important integral part of an overall *correctness by integration* verification strategy. Our methodology provides two benefits: it makes large industrial designs tractable and it incorporates a re-use concept to increase productivity in industrial settings.

References

[1] F. Aglietti, AHB System Generator. http://www.opencores.org/projects.cgi/web/ahb_system_generator/overview, 2004

[2] ARM Limited, AMBA specification (rev 2.0). http://www.arm.com, 1999

[3] A. Biere, A. Cimatti, E. Clarke, Y. Zhu, Symbolic model checking without BDDs, in *Proceedings of the International Conference on Tools and Algorithms for the Construction and Analysis of Systems (TACAS)*, 1999

[4] R.K. Brayton, G.D. Hachtel, A. Sangiovanni-Vincentelli, F. Somenzi, A. Aziz, S.-T. Cheng, S. Edwards, S. Khatri, Y. Kukimoto, A. Pardo, S. Qadeer, R.K. Ranjan, S. Sarwary, T.R. Shiple, G. Swamy, T. Villa, VIS: a system for verification and synthesis, in *Proceedings of the International Conference on Computer-Aided Verification (CAV)*, volume 1102, ed. by R. Alur, T. Henzinger (New Brunswick, NJ, 1996), pp. 428–432

[5] G. Cabodi, S. Nocco, S. Quer, Improving SAT-based bounded model checking by means of BDD-based approximate traversals. J. Univ. Comput. Sci. **10**, 1693–1730 (2004)

[6] E. Clarke, O. Grumberg, S. Jha, Y. Lu, H. Veith, Counterexample-guided abstraction refinement for symbolic model checking. J. ACM **50**(5), 752–794 (2003)

[7] R.P. Kurshan, *Computer-Aided Verification of Coordinating Processes – The Automata-Theoretic Approach* (Princeton University Press, Princeton, NJ, 1994)

[8] M.D. Nguyen, M. Thalmaier, M. Wedler, J. Bormann, D. Stoffel, W. Kunz, Unbounded protocol compliance verification using interval property checking with invariants. IEEE Trans. Comput. Aided Design **27**(11), 2068–2082 (2008)

[9] Onespin Solutions GmbH, Germany. OneSpin 360MV, www.onespin-solutions.com

[10] M. Sheeran, S. Singh, G. Stalmarck, Checking safety properties using induction and a SAT-solver, in *Proceedings of the International Conference on Formal Methods in Computer-Aided Design (FMCAD)*, 2000

[11] K. Shimizu, L. Dill, A.J. Hu, Monitor-based formal specification of PCI, in *Proceedings of the International Conference on Formal Methods in Computer-Aided Design (FMCAD)*, 2000

[12] M. Thalmaier, M.D. Nguyen, M. Wedler, D. Stoffel, W. Kunz, Formale Verifikation von SoC Protokollimplementierungen, in *Tagungsband 1.GMM/GI/ITG-Fachtagung Zuverlaessigkeit und Entwurf*, 2007

[13] C. Wang, B. Li, H. Jin, G. Hachtel, F. Somenzi, Improving Ariadne's bundle by following multiple threads in abstraction refinement. IEEE Trans. Comput. Aided Design **25**, 2297–2316 (2006)

[14] Y.-C. Yang, J.-D. Huang, C.-C. Yen, C.-H. Shih, J.-Y. Jou, Formal compliance verification of interface protocols, in *Proceedings of the IEEE International Symposium on VLSI*, 2005

Chapter 13

ISIS: RUNTIME VERIFICATION OF TLM PLATFORMS

Luca Ferro and Laurence Pierre
TIMA Laboratory (CNRS-GrenobleINP-UJF), 46 Av. Félix Viallet, 38031 Grenoble cedex, France
Luca.Ferro@imag.fr, Laurence.Pierre@imag.fr

Abstract The context of this chapter is the dynamic assertion-based verification (ABV) of TLM SystemC models. We have developed a methodology for checking temporal properties during the SystemC simulation. The assertions are expressed in the PSL language, including the possibility to use its *modeling* layer, and the method supports timed as well as untimed TLM descriptions. It is implemented in a prototype tool called ISIS. We describe its principles and technical characteristics, and we report various experimental results.[1]

Keywords: Assertion-based verification, SystemC TLM, PSL

1. Introduction

With the increasing complexity of modern Systems on Chips (SoC's), the interest in languages like SystemC is growing. SystemC [10] is a library of C++ classes for modeling circuits. TLM (*Transaction Level Modeling*) is much more abstract than RTL (*Register Transfer Level*), it provides communication models for complex data types between the components of the SoC. Compared to RTL, the simulation of TLM models is several orders of magnitude faster, thus considerably improving productivity in SoC design [14]. Hence TLM is widely being adopted, and TLM specifications tend to become golden reference models [7]. In this chapter, we propose a tool for the *assertion-based verification* (ABV) of these models.

In the ABV framework, logic and temporal assertions written in languages like PSL [11] are used to capture the design intent [5, 9]. These assertions are checked using static (model-checking) or dynamic (simulation-based) techniques. This is now a well-established technology *at the RT level*. Even if

D. Borrione (ed.), *Advances in Design Methods from Modeling Languages for Embedded Systems and SoC's*, Lecture Notes in Electrical Engineering 63,
DOI 10.1007/978-90-481-9304-2_13, © Springer Science+Business Media B.V. 2010

various efforts have recently been reported, there is nowadays no simple and effective solution for ABV at the transactional level.

TLM specifications are algorithmic descriptions in which communications are implemented as *function calls*. Time can be introduced, essentially for the purpose of performance evaluation, but there is *no synchronization clock*, and even no notion of cycle-accuracy at the highest abstraction levels. Thus there is a crucial need for ABV solutions that do not rely on discrete time models associated with clocked contexts. Moreover, being able to make use of the *Modeling layer* of PSL is very desirable: this layer enriches PSL by giving the ability to introduce auxiliary variables in the assertions; it is mandatory for expressing most non-trivial properties of TLM models.

In this chapter we describe ISIS, our prototype tool for the verification of temporal properties during SystemC simulation of timed or untimed TLM specifications. It is based on the principles presented in [20], which are extended here to support auxiliary variables managed through the PSL Modeling layer.

2. A Framework for the ABV of TLM Models

The core of the PSL language is the *Temporal layer*, it gives the possibility to describe complex temporal relations, evaluated over a set of evaluation cycles. The *Boolean layer* is used to build basic expressions commonly used by the other layers. The *Modeling layer* is used to augment what is possible using PSL alone, for instance it allows to manage auxiliary (global) variables. The *Boolean* and *Modeling* layers borrow the syntax of the HDL in which the PSL assertions are included. We talk about VHDL, Verilog or SystemC "flavors".

Dynamic verification techniques consider the *simple subset* of PSL, which conforms to the notion of monotonic advancement of time. ISIS supports all *simple subset* formulas of the FL (Foundation Language) class of PSL, which essentially represents the linear temporal logic. The semantics of PSL properties [11] is defined with respect to finite or infinite *execution traces*, or words over the alphabet $\Sigma = 2^{\mathbf{P}} \cup \{\top, \bot\}$ ($\mathbf{P}$ is a non-empty set of atomic propositions, $\top$ and $\bot$ are such that, for every Boolean expression b, $\top \models b$ and $\bot \not\models b$). The length of a trace v is denoted $|v|$ and $v^{i..}$ indicates the suffix of v starting at v^i. In the following, φ and ψ are FL formulas. FL operators are expressed with regard to a small set of basic operators.

One key basic operator is *strong until*, denoted until! or U. Roughly speaking, $\varphi\, U\, \psi$ holds iff there exists an evaluation point from which ψ holds, and φ holds until that point:

$$v \models [\varphi U \psi] \iff \exists k < |v| \; s.t. \; v^{k..} \models \psi \; and \, \forall j < k, v^{j..} \models \varphi$$

The PSL formula always φ means that φ must be verified on each evaluation point of the trace:

$$\text{always } \varphi \stackrel{def}{=} \neg[true\, U\, \neg\varphi]$$

The strong next_event! operator requires the satisfaction of φ the next time a boolean expression b is verified:

$$\text{next\_event!}(b)(\varphi) \stackrel{def}{=} [\neg b\, U\, b \wedge \varphi]$$

Classically, the weak until operator until does not impose the occurrence of ψ. Weak versions of other operators (such as weak next_event) are derived accordingly. See [11] for more details about the PSL operators.

Monitoring TLM Descriptions – Principles

Our methodology enables the dynamic verification of PSL properties for TLM specifications [20]. Two main problems have been solved:

- The automatic construction of efficient monitors (checkers) associated with the PSL properties: instead of using an automata-based approach, the monitors are built as the linear composition of primitive components associated with the elementary PSL operators. To that goal, we have adapted a proven correct methodology for RTL monitors [17]. The primitive components have been defined once and for all and are grouped in a library.
- The definition of a specialized method, summarized below, for monitoring transactional descriptions, in a context where there is no synchronization clock.

We recall that the FL semantics is defined with regard to *execution traces* i.e., words over the alphabet $\Sigma = 2^{\mathbf{P}} \cup \{\top, \bot\}$ where $\mathbf{P}$ is a non-empty set of atomic propositions. At the *synchronous RT level*, the letters of the execution traces correspond to observations that are made on the clock ticks, and the atomic propositions of $\mathbf{P}$ are Boolean expressions over the signals of the system. At the *transactional level*, we leave the PSL semantics unchanged but we consider that observation points are when the assertion needs to be re-evaluated i.e., each time a variable of the assertion is potentially modified (an appropriate event occurs in the corresponding channel). Hence, *all the events that may enable the observation of updated values for the variables involved in the assertion* give rise to new letters in the execution traces. Thus our atomic propositions of $\mathbf{P}$ are SystemC Boolean expressions that express communication events (we mean occurrences of transactions, not SystemC events) and conditions on these communications.

Technically, we employ a model that allows to observe these transactional events and to trigger the monitors when needed [20]. The monitors are enclosed in special *Wrapper* components that analyze the observations and provoke calls to the monitor evaluation functions. For efficiency reasons, the analysis of an assertion A only uses the observation of the subset of events associated with the channels involved in A. Our technology is not restricted to using our checkers, but also allows for upgrading other PSL-oriented checkers to TLM, though those are inherently bound to the RT level. This is feasible provided that they are equipped with a C++ member function that gives their transition function, which is the case with the SystemC checkers generated by *FoCs* [2]. Comparative CPU times will be reported in Section 5.

Discussion About the "Next" Operator

The meaning of the $\mathrm{next!}$ operator is that the sub-property denoted by its operand should be verified from the next evaluation point of the execution trace:

$$v \models \mathrm{next!}\,\varphi \iff |v| > 1 \; and \; v^{1..} \models \varphi$$

As mentioned for instance in [3], with SystemC transactional models (in particular with untimed ones), the execution order may depend on the simulation kernel (the algorithm that determines the order of process activation). Therefore, using the $\mathrm{next!}$ operator for the expression of properties may be unsound. We support it, but we consider that the $\mathrm{next\_event!}$ operator is preferable: it has the advantage of making explicit its expected event, thus avoiding any confusion.

The $\mathrm{next!}$ operator can only be necessary with *inclusive operators*, to exclude the current communication. For instance, the implication operator $\rightarrow$ checks whether its right operand FL property holds *from the moment* its left operand holds (in the PSL "simple subset", the left operand is Boolean). Thus, with the assertion $b \rightarrow f$, f is checked from the moment b holds. If we want to exclude that instant, the assertion must be $b \rightarrow \mathrm{next!}\,f$.

3. Using the Boolean and Modeling Layers

Boolean Layer: C++ Predicates

According to the PSL reference manual, the "SystemC flavor" of the *Boolean layer* should allow the use of any SystemC Boolean expressions. As mentioned in Section 2.0, our atomic propositions are SystemC Boolean expressions that express communication events and conditions on these communications. We give the user the possibility to select the relevant communications for each property. Being able to identify the critical communication events in a given context is crucial. For instance, depending on his confidence in a FIFO channel, the user may prefer to activate the verifications when the FIFO is written or when it is read.

From the set of selected communication operations, we *automatically* generate ad hoc Boolean methods, e.g., `write_CALL()` for *write* actions in a FIFO. The expression `f.write_CALL()`, where `f` is an instance of FIFO, can then be used in assertions to express the occurrence of a *write* action.

Modeling Layer: Auxiliary Variables

To illustrate the necessity of the *Modeling layer*, let us consider the example of a Motion-JPEG decoding platform described at the Transaction Accurate level [6], see Fig. 13.1. It embeds a configurable number of processors (here we use one processor), a global memory, a hardware semaphore RAM, and hardware terminals (TTY). The traffic generator takes its data from a MJPEG file, and RAMDAC is the viewer.

Among the properties that can be verified to check the reliability of the communication channel, the following one needs the Modeling layer: *the data that are written on the RAMDAC are exactly the ones that have been transmitted by the EU*. The expression of this property requires the memorization of the data that are transmitted by the EU, to be compared to the data written on the RAMDAC:

```
/* Modeling layer */
HDL_DECLs:
unsigned int req_data, written_data;
HDL_STMTs:
if (eu.write_CALL()) req_data = eu.write.p2;
if (rdac.write_CALL()) written_data = rdac.write.p2;
/* Assertion */
PROPERTY:
assert ALWAYS (eu.write_CALL() ->
          NEXT_EVENT(rdac.write_CALL())(req_data==written_data));
```

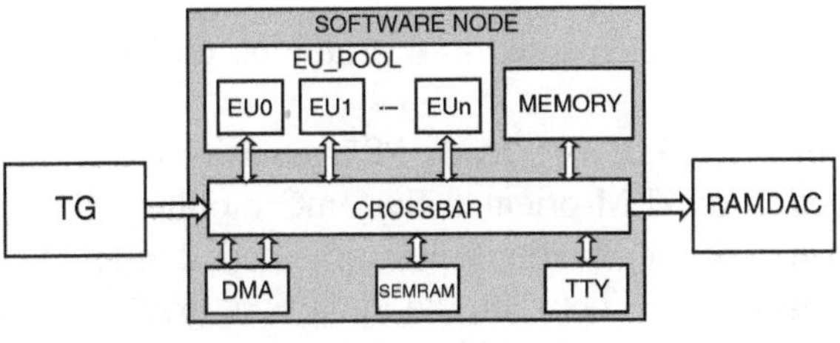

Figure 13.1. M-JPEG platform [6]

In this property, `req_data` is the variable that memorizes the transmitted data; it is updated when `eu` executes its method `write`, we use `eu.write.p2` to denote the second parameter of function `write`. For the sake of clarity, we also use the Modeling layer to manage a variable `written_data` that stores the data written on the RAMDAC. It is not mandatory because its value is checked at the same time it is known (a call to the method `write` of the RAMDAC occurs), therefore `rdac.write.p2` could be used directly in the `NEXT_EVENT` expression. This example is written in the syntax supported by ISIS[2]:

- The variable declarations are introduced by the keyword `HDL_DECLs`. Any legal C++ data type can be used, the variables can also be initialized.
- The statements are introduced by the keyword `HDL_STMTs`. They are given in the C++ syntax.
- The actual definition of the property is introduced by the keyword `PROPERTY`.

During the construction of the monitor, this is interpreted as the association of variables to the property. The corresponding declarations and statements are automatically put in the monitor body, at the beginning of its activation function (thus the statements are executed prior to evaluating the assertion). Anywhere, in the Modeling layer statements or in the assertion, the following syntax can be used:

- `name.fctname_CALL()` denotes that the communication function `fctname` of the element `name` as just been called.
- `name.fctname.p#` denotes the parameter in position # of function `fctname` (0 is used for the return value).

Dealing with additional variables is also addressed for instance in [16]. However, the problem and its solution are significantly different from ours, since they consider the special context of SVA local variables, at the RT level, and the purpose is to generate *hardware* components for *static* verification.

4. The ISIS Tool

The method has been implemented in a prototype tool with a graphical user interface.

As pictured in Fig. 13.2, *ISIS* reads PSL assertions and performs the automatic construction of TLM-oriented SystemC monitors. Then the SystemC code of the design under verification is instrumented, through the GUI, to make explicit the correspondence between the identifiers involved in the properties (except the Modeling layer variables) and actual variables in the design. The monitors are thus linked to the design. This instrumented design is compiled

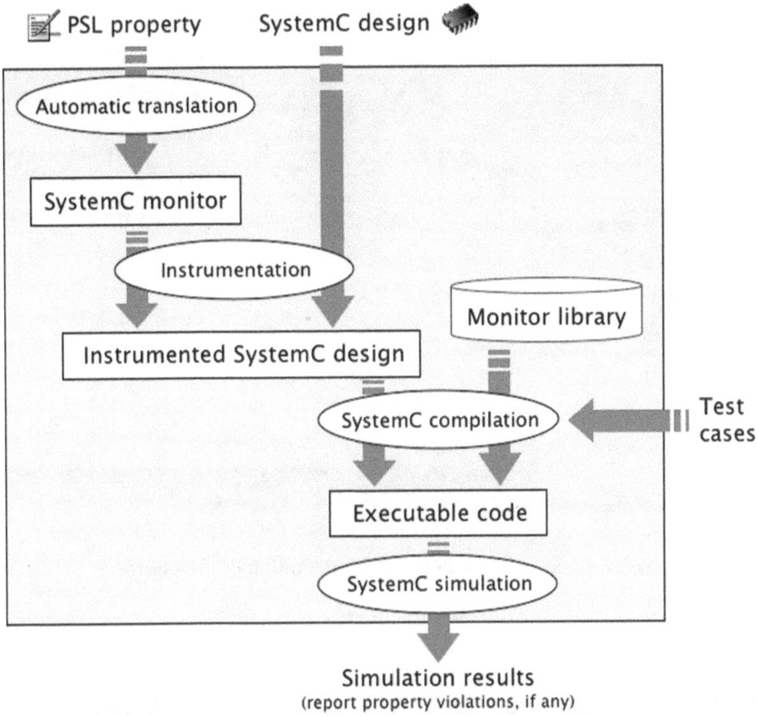

Figure 13.2. ISIS: SystemC designs instrumented for ABV

using our library of primitive monitors, and the SystemC simulator can be run on this combination of modules. Property violations (if any) are reported during simulation.

Figure 13.3 gives a screenshot of *ISIS*, with the example of Section 3.0 (three text areas are proposed for more user-friendliness, the generated code is the one of Section 3.0). The *Properties* tab of the GUI is used to enter the assertions, and the *Interconnect* tab allows to establish the relationship between the identifiers of the property and the actual variable or constant identifiers in the description. In this screenshot, the user has already selected the `write` methods of `eu` and `ramdac` as observed communications, and ISIS has generated the corresponding predicate functions `write_CALL()` (displayed on the right). These functions can be used in the expression of the assertion.

5. Experimental Results

We now report a variety of results that illustrate the capabilities of *ISIS*. In Table 13.1, CPU times are taken on an Intel Core2 Duo (3 GHz) under Debian Linux. The second column gives SystemC simulation times without

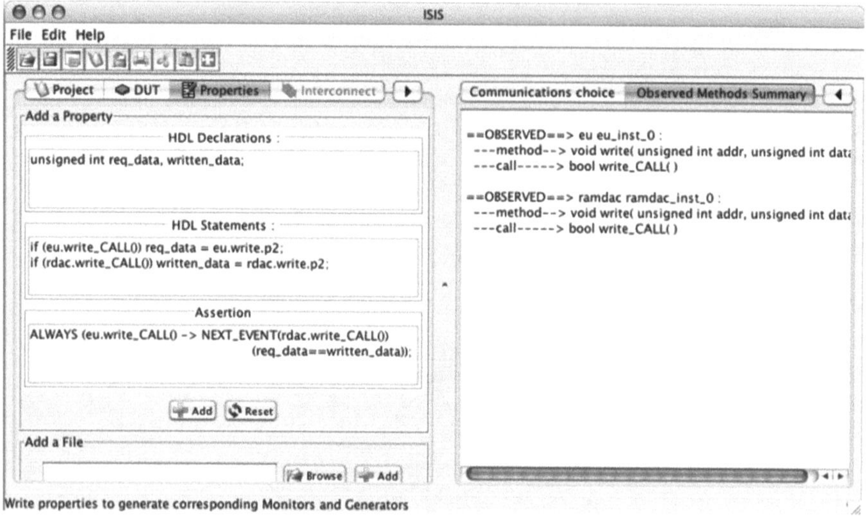

Figure 13.3. Screenshot of ISIS: definition of the assertion

any monitoring. The third and the fourth columns give CPU times for simulation while monitoring with our checkers and with the *FoCs* 2.04 checkers. The number of times the checker functions have been evaluated is reported in the fifth column. A (M) in the first column indicates where the Modeling layer is necessary (*FoCs* does not support the Modeling layer; it is supported by ISIS, in exactly the same way for the *FoCs* and ISIS checkers). Table 13.2 gives CPU times for checking all the properties of each example simultaneously. Checkers construction and code instrumentation times are negligible. For all the examples, simulations have been performed with and without introducing bugs in the design; all property violations have been detected.

Producers/Consumer with a FIFO Channel

In the first example, referred to as FIFO (1) in Table 13.2, a producer sends messages to a consumer through a FIFO channel. The *fifo* class of the SystemC distribution is used. Property P1 is *the consumer does not receive a new message until the previous transfer is completed.* Since we want to verify that the producer behaves as expected by the consumer, but also that the FIFO does not mix consecutive messages, we express this property in terms of the characters *received* by the consumer. The relevant events are thus the acquisitions of data by the consumer i.e., its calls to the *read* method of the FIFO. In the simulations reported in Tables 13.1 and 13.2, the producer sends 10,000 messages of random length (between 1 and 10,000 characters).

We now have two producers that share a FIFO to send messages to a consumer, under the control of an arbiter (FIFO (2) in Table 13.2). Two properties are considered:

- Mutual exclusion between the signals *grant1* and *grant2* of the arbiter (P2).
- *Each time the signal grant1 of the arbiter has a positive edge, then the first client will send its message* (P3).

Two signals are observed to check P2. The signal *grant1* and the *write* operations of the FIFO are observed for the analysis of P3. In the simulations, each producer still sends 10,000 messages of random length (between 1 and 10,000 characters).

DMA System

This is a simple DMA example delivered with TLM 2.0 draft 1. A master programs the DMA through a router in order to perform transfers between two memories:

- It writes into DMA registers the source address, the destination address, and the length of the transferred data.
- Then it requests for the transfer by writing into the DMA control register.

Afterwards, it waits for the *dma_irq* signal from the DMA that indicates the completion of the transfer, and clears the DMA interrupt by resetting the control register. Once the DMA has received all the needed data, it performs the transfer between the two memories by alternating read and write operations. We consider three properties:

- *Any time the control register is programmed, an end-of-transfer notification occurs before the next writing into the control register* (P4).
- *Read and write operations in the memories alternate* (P5).
- *Any time a source address is transferred, a read access in the first memory eventually occurs and the right address is used* (P6).

Test coverage analysis can also easily be performed by the addition of monitors equipped with counters, for instance here to evaluate the repartition of read/write operations in different parts of the memories. We also give CPU times for P6 + coverage analysis.

For this example, 200,000 memory transfers are performed, data of length 64 bytes are transferred (each atomic transfer carries 4 bytes i.e., 16 transfers are needed). For instance, the monitor for P6 is evaluated 4.2 million times (21 times for each transfer: 5 writings to the DMA + 16 read operations).

Protocol over Faulty Channel

This case study is a SystemC version of the protocol over faulty channel described in [1]. A sender sends messages to a receiver through a channel that can lose or duplicate messages. According to the values of the acknowledgment signals, the messages are re-emitted or not. Two properties are verified, that require the Modeling layer:

- *We always have* $ackr \leq nb \leq ackr + 1$, where $ackr$ is the acknowlegment sent by the receiver and nb is the number of the last message sent by the sender (P7).
- *The acknowlegment sent by the receiver is always equal to the length of the actual message it received* (P8).

In the simulations reported in Tables 13.1 and 13.2, the sender sends 4,000 times a random number of messages (between 1 and 4,000).

Table 13.1. Experimental results

	Monitoring			Number of property
	No (s)	ISIS checkers (s)	FoCs checkers (s)	evaluations
P1	4.56	9.70	340.31	50 millions (average)
P2	8.85	8.91	8.93	40,000 (max.)
P3	8.85	17.98	456.38	100 millions (av.)
P4	4.97	5.18	9.80	1.4 millions
P5	4.97	6.21	27.35	6.4 millions
P6 (M)	4.97	5.54	38.88	4.2 millions
P6 + coverage (M)	4.97	6.14	39.15	4.2 + 6.4 millions
P7 (M)	8.97	11.25	12.66	24 millions (av.)
P8 (M)	8.97	10.50	13.21	20 millions (av.)
P9	8.94	11.59	84.65	5 millions (av.)
P10	8.94	9.74	26.92	5 millions (av.)
P11 (M)	18.67	21.64	84.27	13.7 millions (av.)

Table 13.2. Experimental results – simultaneous checks

	Checking all properties simultaneously	
	ISIS (s)	FoCs (s)
FIFO (1)	9.70	340.31
FIFO (2)	17.98	456.38
DMA (P4 + P5 + P6 with coverage)	7.35	65.94
Faulty channel	12.21	16.67
Packet switch	12.13	101.56
MJPEG platform	21.64	84.27

Packet Switch

This is the 4×4 multicast helix packet switch given with the SystemC distribution. This switch uses a self routing ring of shift registers to transfer packets from one port to another in a pipelined fashion. Input and output ports have FIFO buffers of depth four each. Two properties are reported:

- *Absence of starvation* (P9)
- *Absence of deadlock* (P10)

The number of iterations is set to 5 millions. In each iteration, one packet per input and output is processed.

Motion-JPEG Case Study

This is the Motion-JPEG decoding platform of Section 3.0. The property is also the one of Section 3.0: *the data that are written on the RAMDAC are exactly the ones that have been transmitted by the EU* (P11). A simulation that corresponds to 10 seconds of video decoding is reported.

Figure 13.4 gives a screenshot of a simulation while checking property (P11), with design errors introduced in the communication channel. While the image on the RAMDAC is not as expected, our monitor reports detected errors (on the left).

The results of Tables 13.1 and 13.2 demonstrate that our monitoring technology incurs moderate overhead in simulation time. Our monitors outperform

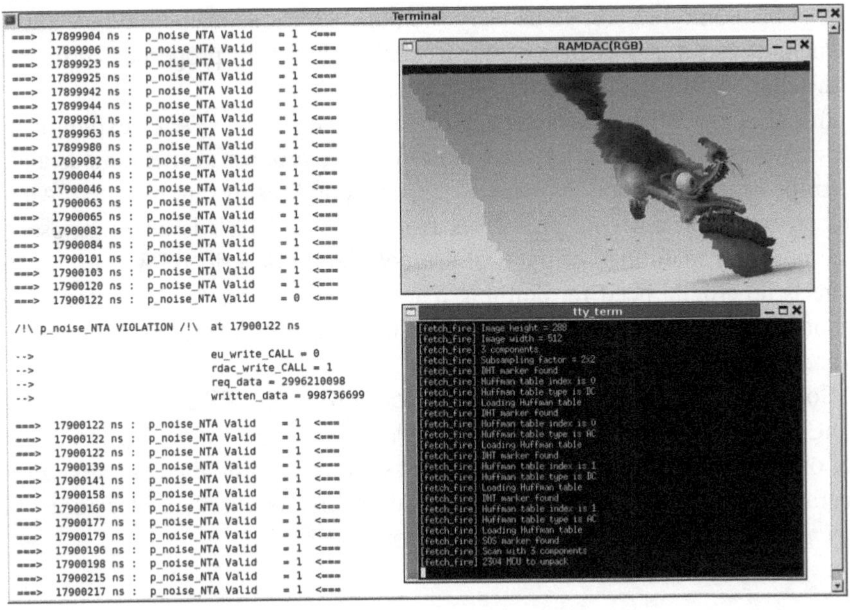

Figure 13.4. Instrumented simulation of the MJPEG platform

the FoCs checkers in most cases. The construction methods are fundamentally different: the *ISIS* monitors are built compositionally, thus their size is linear in the number of PSL operators of the assertion, whereas FoCs uses an automata-based technique. Moreover, it seems that time consumption with FoCs worsens due to the use of complex datatypes.

We can also remark that the cumulative times of Table 13.2 are far from the sum of the times for all the properties. This is essentially due to the fact that, when several properties require the observation of the same channel(s), the observation times are not cumulated. Hence it is worth checking several assertions simultaneously.

6. Related Work

The approaches proposed in [8, 12, 18] and concentrate on *model checking* SystemC designs. They consider either clock-synchronized descriptions [8] or actual TLM specifications [12, 18]. Abstraction techniques are required to get tractable models, or only limited (pieces of) designs can be processed.

A *simulation-based approach* is proposed in [19]. This approach is very restrictive (it can be applied only to certain categories of TLM descriptions), it requires two simulation runs, and does not allow to take into account the parameter values of transactions.

C++ checkers are generated using *FoCs* in [15]. Specific channel classes are defined to be able to trigger them: a channel can contain instance(s) of checker(s), their transition functions are called when transactions are initiated in this channel. The main limitations of this method are that several channels cannot be monitored simultaneously (i.e., the assertions cannot involve several channels), and that it is impossible to deal with hybrid properties (properties that include both signals and TLM channels).

In [3], a specialized language is defined. It extends SVA to explicitly express events. A possible implementation using proxies is mentioned, and is described more precisely in [4]. Few experimental results are reported. An approach based on deriving primitives from SVA and PSL is also proposed in [13] but the capabilities of the tool are not clearly described. We believe that there is no crucial need for defining a specific language. As illustrated in this chapter, the PSL language, in particular its *Temporal* layer, is rich enough to express a large range of properties.

Some commercial tools also provide for introducing PSL assertions in SystemC designs. The tool *Cadence Incisive Unified Simulator* supports TLM, but only signals can be involved in the assertions, and neither the next_event operator nor the strong temporal operators can be used. Other tools like *Synopsys VCS* only accept RTL descriptions.

7. Conclusion

We have presented our solution for verifying logic and temporal properties about communications in TLM designs during simulation. The tool is usable with timed or untimed TLM models, it supports the introduction of auxiliary variables, properties can mix various types of channels. The technique is also compatible with GALS and multi-clocked systems. Our ABV solution has many advantages:

- It is *based on PSL*, which is an IEEE standard that is nowadays well-established in the CAD community.
- It is *slightly intrusive* in the SystemC code, which is augmented with mechanically generated new classes, and undergoes few modifications in its declarative parts.
- It provides a *high level of automation*: the checkers are automatically generated, and the code is mechanically instrumented with user-guidance through a GUI.
- *Performance results* are excellent.

Future works include investigating the extension of this methodology to the SERE operators of PSL, and relating properties at TLM with corresponding properties at RTL.

Notes

1. This work is partly supported by the French projects SFINCS (ANR) and SoCKET (FCE).

2. The grammar given in the PSL reference manual only specifies that a verification unit can contain HDL declarations and statements as well as assertions. In order to avoid any confusion during parsing, we have chosen special keywords to introduce declarations and statements.

References

[1] K.M. Chandy, J. Misra, *Parallel Program Design: A Foundation* (Addison-Wesley, Boston, MA, 1988)

[2] A. Dahan, D. Geist, L. Gluhovsky, D. Pidan, G. Shapir, Y. Wolfsthal, L. Benalycherif, R. Kamdem, Y. Lahbib, Combining system level modeling with assertion based verification, in *Proceedings of the ISQED'2005*, 2005

[3] W. Ecker, V. Esen, M. Hull, Execution semantics and formalisms for multi-abstraction TLM assertions, in *Proceedings of the MEMOCODE'06*, 2006

[4] W. Ecker, V. Esen, M. Hull, Implementation of a transaction level assertion framework in SystemC, in *Proceedings of the DATE'07*, 2007

[5] H. Foster, A. Krolnik, D. Lacey, *Assertion-Based Design* (Kluwer, Dordrecht (NL), 2003)

[6] P. Gerin, X. Guérin, F. Pétrot, Efficient implementation of native software simulation for MPSoC, in *Proceedings of the DATE'08*, 2008, pp. 676–681

[7] R. Goering, Transaction models offer new deal for EDA. EETimes, 2006, `http://www.eetimes.com/showArticle.jhtml?articleID=181503693`

[8] A. Habibi, S. Tahar, Design and verification of SystemC transaction level models. IEEE Trans. VLSI Syst. **14**(1) (2006), 57–68

[9] J. Horgan, Assertion based verification. EDACafe Weekly, 2004, `http://www10.edacafe.com/nbc/articles/view_weekly.php?articleid=209195`

[10] *IEEE Std 1666–2005, IEEE Standard SystemC Language Reference Manual*. IEEE, 2005

[11] *IEEE Std 1850–2005, IEEE Standard for Property Specification Language (PSL)*. IEEE, 2005

[12] D. Karlsson, P. Eles, Z. Peng, Formal verification of SystemC designs using a Petri-net based representation, in *Proceedings of the DATE'2006*, 2006

[13] A. Kasuya, T. Tesfaye, Verification methodologies in a TLM-to-RTL design flow, in *Proceedings of the DAC'07*, 2007

[14] W. Klingauf, M. Burton, R. Günzel, U. Golze, Why we need standards for transaction-level modeling. SOC Central, 2007

[15] Y. Lahbib, *Extension of assertion-based verification approaches for the verification of SystemC SoC models*. PhD thesis, University of Monastir (Tunisia), 2006

[16] J. Long, A. Seawright, Synthesizing SVA local variables for formal verification, in *Proceedings of the DAC'07*, 2007

[17] K. Morin-Allory, D. Borrione, Proven correct monitors from PSL specifications, in *Proceedings of the DATE'2006*, 2006

[18] M. Moy, F. Maraninchi, L. Maillet-Contoz, LusSy: an open tool for the analysis of systems-on-a-chip at the transaction level, in *Design Automation for Embedded Systems*, 2006

[19] B. Niemann, C. Haubelt, Assertion-based verification of transaction level models, in *Proceedings of the ITG/GI/GMM Workshop*, 2006

[20] L. Pierre, L. Ferro, A tractable and fast method for monitoring SystemC TLM specifications. IEEE Trans. Comput. **57**(10) (2008), 1346–1356

Chapter 14

SMT-BASED STIMULI GENERATION IN THE SYSTEMC VERIFICATION LIBRARY*

Robert Wille, Daniel Große, Finn Haedicke, and Rolf Drechsler
Institute of Computer Science, University of Bremen, 28359 Bremen, Germany
{ rwille,grosse,finn,drechsle } @informatik.uni-bremen.de

Abstract Modelling at the Electronic System Level (ESL) is the established approach of the major System-on-chip (SoC) companies. While in the past ESL design covered design methodologies only, today also verification and debugging is included. To improve the verification process, testbench automation has been introduced highlighted as constraint-based random simulation. In SystemC – the de facto standard modelling language for ESL – constraint-based random simulation is available through the SystemC Verification (SCV) library. However, the underlying constraint-solver is based on Binary Decision Diagrams (BDDs) and hence suffers from memory problems.

In this chapter, we propose the integration of new techniques for stimuli generation based on Satisfiability Modulo Theories (SMT). Since SMT solvers are designed to determine a single satisfying solution only, several strategies are proposed forcing the solver to generate more than one stimuli from different parts of the search space. Experiments demonstrate the advantage of the proposed approach and the developed strategies in comparison to the original BDD-based method.

Keywords: Constraint-based random simulation, SAT modulo theories, SystemC verification library

1. Introduction

The *Electronic System Level* (ESL) emerged to deal with both, the (ever increasing) design complexity and the resulting verification crisis. Hence, the languages for ESL design, e.g. SystemVerilog or SystemC, have been

*This work was supported in part by the German Federal Ministry of Education and Research (BMBF) within the project SANITAS under contract no. 01M3088.

D. Borrione (ed.), *Advances in Design Methods from Modeling Languages for Embedded Systems and SoC's*, Lecture Notes in Electrical Engineering 63,
DOI 10.1007/978-90-481-9304-2_14,

enhanced to support inevitable verification methodologies like assertions and testbench automation. Especially the second methodology, highlighted as *constraint-based random simulation*, is a key technique to cope with current and future designs [2, 28, 29]. Here, stimuli are generated from specified constraints by means of a constraint solver so that corner case scenarios are triggered. In doing so, design bugs will be found that might otherwise remain undetected.

In the context of SystemC – the de facto standard modelling language for ESL – constraint-based random simulation is available in the *SystemC Verification* (SCV) library. Here, the underlying constraint-solver uses *Binary Decision Diagrams* (BDDs) [5] for constraint representation and subsequent stimuli generation. Due to the well-known limits of BDDs, complex arithmetic or non-trivial implications can not efficiently be handled by SCV constraints.

In this work, we introduce an alternative to BDD-based stimuli generation. Motivated by the solving paradigms introduced in the recent past, the BDD-based constraint solver is replaced by a *Satisfiability Modulo Theory* (SMT) approach. SMT extends *Boolean Satisfiabily* (SAT), a solving technique on the pure Boolean level that has been intensely studied over the last two decades. SMT maintains all benefits of SAT but furthermore allows the exploitation of additional information due to the representation of the instances on the word level. This often gives significant speed-ups compared to pure SAT solving (see e.g. [1, 18, 31]).

While, at a first glance the substitution of BDDs by SMT techniques seems to be simple, two major tasks have to be tackled:

1 BDDs represent all solutions of a constraint at once. Hence, if the BDD can be built, stimuli generation is nothing else than a linear traversal from the root node to a 1-terminal respecting a uniform distribution. In contrast, SMT solvers return a single solution only. Thus, strategies have to be developed to determine more than one stimuli with good distributions.

2 For SAT solving, MiniSat [10] is the state-of-the-art solver today. However, in the domain of SMT the results are not so clear as e.g. the recent SMT competition shows [24]. Moreover, most of the SMT solvers (in particular the most efficient ones) are closed source. Thus, the proposed strategies for determining good distributions can only use the provided interface of an SMT solver restricting the strategies to a certain extend.

To the best of our knowledge SMT-based stimuli generation has not been considered so far. In [16], an SMT-like static analysis is used to reduce the search space before applying SAT- or BDD-based stimuli generation. Besides that, only approaches exploiting SAT techniques have been proposed. Thereby, similar issues must be addressed. In [21], a specialized SAT algorithm in combination with so called randomized XOR constraints are used to achieve even distributed solutions. Instead, the authors in [17] presented a sampling algorithm that combines concepts from the Metropolis-Hastings algorithm, Gibbs sampling, and WalkSAT to efficiently generate solutions to the constraints with an approximately uniform distribution. In our work, we propose a variety of strategies (including special blocking constraints, solution selection, and handling overconstraining) considering the particular needs of currently available SMT solvers.

The remainder of this chapter is structured as follows: The next section briefly describes the SCV library, BDD-based constraint solving, and the main concepts of SMT. After discussing the limits of BDD-based constraint solving our SMT-based approach is introduced in Section 3. Section 4 describes strategies to generate distributed solutions. Finally, experimental results as well as conclusions with outlooks on future work are given in Section 5 and Section 6, respectively.

2. Preliminaries

To keep the remainder of this paper self-contained, this section briefly introduces the SystemC Verification Library which enables constraint-based random simulation for SystemC. Afterwards, SMT and the respective solve engines are reviewed. Readers wishing more in-depth treatment are referred to the references.

SystemC Verification Library

The SCV library was introduced in 2002 as an open source C++ class library [15, 22, 23] on top of SystemC [14, 26]. In the following we focus only on the basic features of the SCV library for constraint-based randomization.

Using the SCV library, constraints are modelled in terms of C++ classes. That way constraints can be hierarchically layered using C++ class inheritance. In detail, a constraint is derived from the `scv_constraint_base` class. The data to be randomized is specified as `scv_smart_ptr` variables.

EXAMPLE 14.1 *Figure 14.1 shows an example of an SCV constraint with the name* cstr. *Here, the three unsigned integer variables a, b, and addr are randomized. The conditions on the variables a, b, and addr are defined by expressions in the respective* SCV_CONSTRAINT() *macro.*

Internally, a constraint in the SCV library is given by the corresponding characteristic function, i.e. the function is true for all solutions of the constraint. This characteristic function of a constraint is represented as a BDD, a canonical and compact data structure for Boolean functions [5]. For stimuli generation a weighting algorithm is applied to the constraint BDD to guarantee a uniform

```
struct cstr : public scv_constraint_base {
    scv_smart_ptr<sc_uint<64> > a,b,addr;
    SCV_CONSTRAINT_CTOR(cstr) {
        SCV_CONSTRAINT( a()  > 100 );
        SCV_CONSTRAINT( b() ==   0 );
        SCV_CONSTRAINT( addr()>=0 && addr()<=0x400 );
    }
};
```

Figure 14.1. Example constraint

distribution of all constraint solutions and hence maximizing the possibility for entering unexplored regions of the design state space (see [23] and for improvements see [12, 13]). As BDD package CUDD [25] is used in the SCV library.

SAT Modulo Theory

SAT Modulo Theory (SMT) is an extension of the *Boolean satisfiabiliy* problem (SAT problem) which is defined as follows:

DEFINITION 14.2 *Let f be a Boolean function. Then, the SAT problem is to determine an assignment α to all variables of f such that $f(\alpha) = 1$ or to prove that no such assignment exists. In the former case f is* satisfiable; *otherwise f is* unsatisfiable.

Usually, the Boolean function f is given in *Conjunctive Normal Form* (CNF), i.e. a product-of-sum representation. For this representation, several (backtracking) algorithms (so called *SAT solvers*) have been proposed in the past (e.g. [7, 10, 19, 20]). Most of them are based on three essential procedures: (1) the decision heuristic assigns values to free variables, (2) the propagation procedure determines implications due to the last assignment(s), and (3) the conflict analysis tries to resolve conflicts by backtracking that occur during the search. Thereby, advanced techniques as e.g. *efficient Boolean constraint propagation* [20] or *conflict analysis* [19] are exploited. These enable the consideration of problems with more than hundreds of thousands of variables and clauses. Thus, today SAT solvers are core solve engines for many application domains. Thereby, the real world problem is transformed into CNF and then solved by using a SAT solver as black box.

However, due to the increasing complexity of many problems, researchers investigated the use of higher levels of abstractions than CNF – by still exploiting the established SAT techniques. This leads to the development of SMT solvers. Here, the satisfiability problem is expressed e.g. by *Quantifier-Free Bit-Vector* (QF_BV) logic which is defined as follows:

DEFINITION 14.3 *A* bit-vector *is an element* $\vec{b} = (b_{n-1}, \ldots, b_0) \in \mathbb{B}^n$. *The* index $[\,] : \mathbb{B}^n \times [0, n) \to \mathbb{B}$ *maps a bit-vector* $\vec{b}$ *and an index* i *to the* i^{th} *component of the vector, i.e.* $\vec{b}[i] = b_i$. *The conversion from (to) a natural number is defined by* $\text{nat} : \mathbb{B}^n \to N$ *(*$\text{bv} : N \to \mathbb{B}^n$*) with* $N = [0, 2^n) \subset \mathbb{N}$ *and* $\text{nat}(\vec{b}) := \Sigma_{i=0}^{n-1} b_i \cdot 2^i$ *(*$\text{bv} := \text{nat}^{-1}$*).*

Problems can be specified by using bit-vector operations as well as arithmetic operations. Let $\vec{a}, \vec{b} \in \mathbb{B}^n$ *be two bit-vectors. Then, the* bit-vector operation $\circ \in \{\wedge, \vee, \ldots\}$ *is defined by* $\vec{a} \circ \vec{b} := (\vec{a}[n-1] \circ \vec{b}[n-1], \ldots, \vec{a}[0] \circ \vec{b}[0])$. *An* arithmetic operation $\bullet \in \{*, +, \ldots\}$ *is defined by* $\vec{a} \bullet \vec{b} := \text{bv}(\text{nat}(\vec{a}) \bullet \text{nat}(\vec{b}))$. *For brevity, in the following* $\text{nat}(\vec{a})$ *is also written as* a.

Satisfiability problems in QF_BV logic are solved either by (1) using a combination of a traditional SAT solver and a specialized (bit-vector) theory solver (see e.g. [3, 9]), (2) pre-processing the instance exploiting the higher level of abstraction before bitblasting it to a traditional SAT solver (see e.g. [4, 11]), or (3) using specialized solvers that directly work on the bit-level of the problem (e.g. [8, 27]).

3. SMT-Based Stimuli Generation

This section introduces the SMT-based approach for constraint-based stimuli generation. As described in Section 2, the current constraint-solver of the SCV library uses BDDs. Limits of this method are discussed in the first part of this section. Then, we introduce the SMT approach which replaces the BDD-based constraint-solver. Using this as a basis, strategies to determine good distributions are given in the next section.

Limits of BDDs

In the SCV library, constraints are represented by means of BDDs. This allows a compact representation of the respective characteristic function. Furthermore, a uniform distribution among all possible stimuli is ensured.

The flow of the BDD-based stimuli generation is depicted in the upper part of Fig. 14.2. By instantiating the SCV constraint, an expression tree of the constraint is generated. Afterwards, this tree is traversed to build the respective BDD. Stimuli are obtained by randomly selecting a weighted 1-path of the BDD which represents a satisfying assignment to the constraint [30]. Using the improvements for the SCV library as described in [12] a uniform distribution results, even if constraint variables are fixed to certain values.

However, even though BDDs are very successful as a compact data-structure for large Boolean functions, their capabilities are limited. More precisely, for

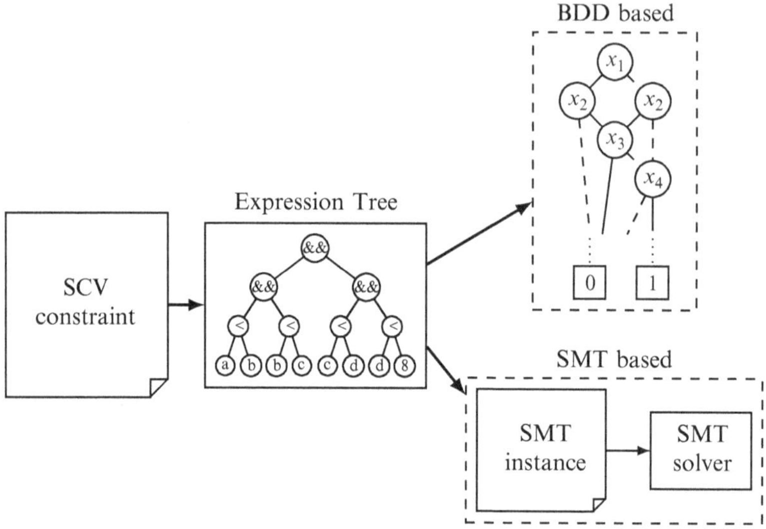

Figure 14.2. Flows in the SCV library

many (practical relevant) functions no compact BDD can be built. As an example, it has been proven, that BDDs representing multiplication always have exponential size with respect to the number of input variables regardless of the variable ordering [6]. Since arithmetic operations like multiplication often occur in SCV constraints, this becomes the bottleneck of stimuli generation for many systems.

Besides that, also the above mentioned uniform distribution of stimuli is often sub-optimal. In many cases a uniform distribution is desirable not with respect to all solutions, but with respect to the control parts of the device under verification. For example, if stimuli for an *Arithmetic Logic Unit* (ALU) are generated, each ALU-operation should be tested by the same number of stimuli. However, using the current BDD-based SCV implementation this is not possible (see also Section 5).

Integrating SAT Modulo Theories

To overcome the mentioned limitations (in particular the capability problems), SMT-based constraint solving is proposed in this section. To this end, the BDD engine is replaced by an off-the-shelf SMT solver as depicted in the lower part of Fig. 14.2. This requires the SCV constraint to be encoded as an SMT instance first.

Table 14.1 shows the operations available in the SCV library as well as their counterparts in QF_BV logic. As can be seen, most of the SCV operations can be represented directly by QF_BV operations. The same holds for most of the

Table 14.1. SCV to SMT (QF_BV) mapping

SCV	*SMT*
EQUAL	=
NOT_EQUAL	not (=)
GREATER_THAN	bvugt,bvsgt
LESS_THAN	bvult,bvslt
GREATER_OR_EQUAL	bvuge,bvsge
LESS_OR_EQUAL	bvule,bvsle
AND	and
OR	or
NOT	not
PLUS	bvadd
MINUS	bvsub
MULTIPLY	bvmul
BITNOT*	bvnot
BITAND*	bvand
BITOR*	bvor
BITXOR*	bvxor
BITSELECT*	extract
BITSLICE*	extract

*Extensions from [12].

variable definitions that can be converted to bit-vector variables as defined in Definition 14.3. In doing so, the SCV constraint can be easily encoded as an SMT instance which is afterwards passed to the respective solver. In contrast to BDDs, this lifts the solve process from the Boolean level to a higher abstraction, where the original operations (in particular arithmetic) are preserved and thus the solve engine can exploit more information.

As a drawback, SMT solvers are designed to determine a single satisfying solution, i.e. only one stimulus is generated for each instance. To generate more than one stimulus, solutions must be blocked, otherwise the same solution is found again due to the deterministic behavior of the SMT solver. Therefore, new constraints excluding already found solutions from the solution space are added. Then, the solver is restarted which leads to a new solution and stimulus, respectively. This is repeated until either the desired number of stimuli have been generated or the instance becomes unsatisfiable (i.e. all solutions have been determined).

EXAMPLE 14.4 *Consider the SCV constraint given in Fig. 14.1. Encoding this into an SMT instance and solving it with an SMT solver may lead to the first stimulus* $a = 101$, $b = 0$, *and* $addr = 1$. *To block this solution, an additional SCV constraint stating* $a \neq 101 \vee b \neq 0 \vee addr \neq 1$ *is added. Thus, a new valid solution could be* $a = 102$, $b = 0$, *and* $addr = 1$.

4. Distribution of Generated Stimuli

Performing SMT-based stimuli generation as described in Section 3 leads to sub-optimal distributions of generated stimuli. More precisely, due to the blocking of previously found solutions two consecutive stimuli may only slightly differ, e.g. in one variable (see Example 14.4). In contrast, solutions from a significant different part of the search space or triggering a new operation of the device under verification are desired. In this section, we introduce strategies for SMT-based stimuli generation resulting in well-distributed solutions. Thereby, specialized types of blocking constraints, random pre-assignments, as well as utilization of control variables are applied. Furthermore, strategies for selecting the obtained solutions as stimuli are presented. Finally, we show how the case of overconstraining (caused by an iterative use of the techniques) can be handled.

Determine Distributed Solutions

To achieve good distributions of stimuli, the solution space of the respective SMT instance must be restricted, so that the solver is forced to determine solutions from different areas of the search space. For this task, three strategies are introduced in the following.

The first one uses blocking constraints as already introduced in Section 3. In total, four variations of blocking constraints are proposed:

- *Total Blocking*
  The complete assignments to previously found solutions are blocked. This is equal to the blocking constraints introduced in the last section and ensures, that the same assignments will not be determined by the solver again. As a result, a very small part of the solution space is blocked and further solutions might be quite similar to the already found ones (i.e. the overall distribution might become sub-optimal).

- *Partial Blocking*
  Only partial assignments of the previously found solutions are considered, i.e. some variable assignments are blocked while others remain unrestricted. As a result, a significantly larger part of the search space (the 'surroundings' of the already found solutions) is excluded. This forces the solver to find more different stimuli but may lead to an early overconstraining (i.e. an exclusion of *all* remaining solutions).

- *Interval Blocking*
  Interval blocking is a special case of partial blocking, where – based on the already found solutions – intervals are excluded from the search space.

- *Bitmask Blocking*
  Some bits of the assignments to previously found solutions are blocked. To this end, a random bitmask is generated and compared to the variable assignments. Where the bitmask is one the respective assignment is blocked.

EXAMPLE 14.5 *Again, consider the SCV constraint given in Fig. 14.1. The following constraints can be added to generate distributed solutions:*
Total Blocking: $a \neq 101 \vee b \neq 0 \vee addr \neq 1$
Partial Blocking: $a \neq 101$
Interval Blocking: $a < 50 \vee a > 150 \vee b < 0 \vee b > 50 \vee addr \leq 0 \vee addr > 50$
Bitmask Blocking: $\vec{a}[57] \neq 0 \wedge \vec{a}[52] \neq 0 \wedge \cdots \wedge \vec{a}[0] \neq 1 \wedge \vec{b}[61] \neq 0 \wedge \cdots$

As second strategy, variables are (partially) pre-assigned to direct the SMT solver into other parts of the search space. Thereby, random values are used and applied to randomly chosen bits of the SCV variables. In doing so, widespreaded solutions may result, but the possibility of overconstraining increases since a pre-assignment can be applied for which no satisfying solution exists any longer.

Finally, the third strategy exploits information of control and data paths of the device under verification. Here, in addition to the SCV constraint, the verification engineer also provides whether an SCV variable belongs to the control path (e.g. the variable controls the operation of an ALU). Then, to obtain well-distributed solutions with respect to all possible operations of the device under verification, the following strategies can be applied:

- *Force a Change*
  Two consecutive solutions include different assignments to the control variables. This ensures, that consecutive stimuli are generated, triggering the simulation of two different operations.

- *Assign Control Signals*
  Pre-assigning the control variables to (random) values. This ensures, that different (random) operations are simulated.

EXAMPLE 14.6 *Consider the SCV constraint in Fig. 14.3, which is used as basis for stimuli generation of an ALU. The variable op is marked as control variable that selects the respective operation to be performed.*

Force a Change: *Let* $op = 1$, $a = 411$, $b = 31$ *be a determined solution. Then, the constraint* $op \neq 1$ *is added for the next stimulus generation.*

Assign Control Signals: *Since in total there are four possible assignments to op, for each stimulus generation run, one of the constraints* $op = 0$, $op = 1$, $op = 2$, *or* $op = 3$ *is added, respectively, while the previous constraint is removed.*

```
struct ALU : public scv_constraint_base {
    scv_smart_ptr<sc_bv<2> > op; // control
    scv_smart_ptr<sc_uint<12> > a, b;

    SCV_CONSTRAINT_CTOR(ALU12) {
        SCV_CONSTRAINT ( (op() != 0x0) ||
            ( 4095 >= a() + b() ) );
        SCV_CONSTRAINT ( (op() != 0x1) ||
            ((4095 >= a() - b())
            && (b() <= a()) ) );
        SCV_CONSTRAINT ( (op() != 0x2) ||
            ( 4095 >= a() * b() ) );
        SCV_CONSTRAINT ( (op() != 0x3) ||
            ( b() != 0 ) );
    }
};
```

Figure 14.3. Example constraint

Choosing Solutions

The strategies introduced so far help to obtain different solutions. Besides that, how to select solutions as stimuli has an effect on the resulting distributions. To this end, the following methods are proposed:

- *Skip Solutions*
  Instead of using each determined solution a number of solutions can be skipped. Because the skipped stimuli are still generated – but not used – the applied strategy still generates constraints for these. This can help traversing different parts of the solution space.

- *Consider All Solutions*
  The best results regarding the distribution can be achieved by determining all solutions. Then, each solution can be selected with equal probability. On the other hand, this obviously requires a large amount of run-time and memory, respectively. Thus, this is only possible for small solution spaces.

- *Apply FIFO*
  A trade-off to the consideration of all solutions is the application of a FIFO structure. In this structure, a certain number of solutions is stored. Each time a stimulus is selected, older solutions are replaced by new ones.

Handling Overconstraining

The proposed strategies constrain the search space in different ways. Some strategies are very restrictive so that an overconstraining may occur very fast, i.e. constraints are added such that the resulting SMT instance becomes unsatisfiable. To handle this, some of the additional constraints must be removed. Therefore, the following strategies are proposed:

- *Remove All Random Constraints*
  Remove all constraints including random elements (e.g. random pre-assignments or bitmasks). This may lead to duplications.

- *Remove Old Constraints (FIFO)*
  Remove the oldest constraint that have been added. If then the instance is still unsatisfiable, remove the next-oldest element until the overconstraining is solved. For this purpose, a FIFO structure is used to store the constraints.

5. Experimental Evaluation

The SMT-based stimuli generator as well as the strategies for determining distributed solutions as introduced in Section 3 and Section 4, respectively, have been implemented in C++ and integrated into the SCV library. In the following we give a comparison of the proposed techniques (using Boolector [4] as SMT solver) and the original BDD-based SCV library.

The following SCV constraints have been used to demonstrate the effects[1]:

- *bv256_neq*: A comparison between two 256-bit variables

- *add*: An addition of two 32-bit variables where the sum is forced to be equal to 999999

- *mult32*: A multiplication of two 32-bit variables where no further restrictions are applied (*_abc*) and where the product is forced to be within the interval $[1027, 1072693248]$ (*_uplow*), respectively

- *ALUXX*: An ALU constraint as the one depicted in Fig. 14.3 where *XX* denotes the bit-width of the data variables

For measuring the distribution of the solutions the normalized Hamming distance (*ham*) has been used. More precisely, the sum of the hamming distances for all generated stimuli divided by the number of all distances is applied (i.e. the higher the value of the normalized hamming distance, the better the distribution). Besides that, also the number of duplicates (*dupl*) is considered to evaluate the quality of the distribution.

In our experiments, for each SCV constraint 1,000 stimuli are generated using the following methods:

- BDD: The original SCV environment based on BDDs
- SMT-TB: The naive SMT-based stimuli generation using total blocking (TB) of previous solutions
- SMT-BB_OR: The SMT-based stimuli generation using bitmask blocking (BB) to avoid previously found solutions; in case of overconstraining, the oldest blocking constraint is removed (OR)
- SMT-BB_OR_F64,4: The SMT-BB_OR strategy (see above) combined with a FIFO storing 64 solutions from which four elements are replaced each time (F64,4)
- SMT-INT: The SMT-based stimuli generation using interval constraints (INT) to avoid previously found solutions
- SMT-INT_OR: The SMT-based stimuli generation using interval constraints (INT) to block previously found solutions; in case of overconstraining, the oldest blocking constraint is removed (OR)
- SMT-PB_OR: The SMT-based stimuli generation using partial blocking (PB) to avoid previously found solutions; in case of overconstraining, the oldest blocking constraint is removed (OR)
- SMT-TB_PAV: The SMT-based stimuli generation using total blocking (TB) and (partially) pre-assigned variables (PAV) to avoid previously found solutions
- SMT-TB_PAV_CC: The SMT-TB_PAV strategy (see above) combined with forcing changes on control variables (CC)
- SMT-TB_PAV_PAC: The SMT-TB_PAV strategy (see above) combined with pre-assigning of control variables (PAC)

In the experiments, for each benchmark a timeout of 900 CPU seconds has been set using an AMD Athlon 3500+ with 1 GB of main memory.

The resulting run-times (in CPU-seconds) and the information for the distribution (normalized hamming distance and number of duplications) are given in Tables 14.2 and 14.3, respectively. The best results are thereby highlighted in bold. In case of timeout (denoted by *) or overconstraining (denoted by †) only the number of stimuli generated so far are shown in Table 14.2. Note that, to compare the quality of the approaches, both tables (i.e. run-time *and* distribution) must be considered together.

Table 14.2. Runtimes (CPU seconds)

	bv256_neq	add	mult32_abc	mult32_uplow	ALU12	ALU16	ALU32
BDD	**0.00**	**0.00**	(*0)	(*0)	33.96	(*0)	(*0)
SMT-TB	720.40	90.13	437.15	**271.12**	5.57	5.91	5.94
SMT-BB_OR_F64.4	(*227)	(*630)	(*161)	(*219)	97.52	(*882)	(*579)
SMT-INT	(*559)	(†3)	(*562)	(†36)	(†523)	(†268)	(*973)
SMT-BB_OR	(*950)	110.43	(*719)	(*830)	6.80	6.72	7.64
SMT-INT_OR	(*580)	10.75	(*618)	362.74	8.24	6.37	7.20
SMT-PB_OR	312.83	54.76	419.85	320.53	**3.49**	**3.35**	**2.61**
SMT-TB_PAV	356.96	282.12	**121.84**	288.33	6.35	6.10	6.08
SMT-TB_PAV_CC	351.66	280.50	**122.68**	292.40	7.07	6.89	7.41
SMT-TB_PAV_PAC	345.59	270.89	**121.50**	288.64	7.71	7.31	7.16

In case of overconstraining (denoted by †) or timeout (denoted by *), respectively, the number of generated stimuli so far is given instead.

As can be seen, the original SCV environment (BDD) easily handles *bv256_neq* and add but suffers from time outs as soon as multipliers occur in the respective SCV constraint. In contrast, the naive SMT-based approach (SMT-TB) is able to generate the desired 1,000 stimuli for all benchmarks in reasonable time (see first two rows of Table 14.2). However, the distribution obtained by the SMT-based approach is poor (see first two rows of Table 14.3). This emphasizes the need for advanced strategies as proposed in Section 4.

But not all developed strategies lead to the desired results (i.e. fast stimuli generation *and* good distribution). More precisely, SMT-BB_OR_F64,4, SMT-INT, SMT-BB_OR, and SMT-INT_OR are not applicable since they cannot complete all benchmarks within the timeout. Furthermore, SMT-PB_OR is not suitable due to the high number of generated duplicates (e.g. for *ALU12* 696 of the 1,000 generated stimuli are duplicates). Thus, all approaches listed in rows 3–7 of Tables 14.2 and 14.3, respectively, are not applicable due to the high run-time or low quality of the distribution.

In contrast, the remaining strategies SMT-TB_PAV, SMT-TB_PAV_CC, and SMT-TB_PAV_PAC (lower rows of Tables 14.2 and 14.3) offer the best trade-off between run-time and distribution. All of them directly exclude previous found solutions and thus avoid duplicates. Furthermore, the hamming distance is significantly higher than for the naive SMT approach (SMT-TB) while the run-time remains moderate.

Finally, the effect of the strategies that incorporate control path information are considered in detail. This is done by considering the example of *ALU12*.

Table 14.3. Distribution (normalized Hamming distance, number of duplicates)

	bv256_neq		add		mult32_abc		mult32_uplow		ALU12		ALU16		ALU32	
	ham	dupl	ham	dupl	ham	dupl	ham	dupl	ham	dupl	ham	dupl	ham	dupl
BDD	**127.70**	0	10.01	0	–	–	–	–	**6.42**	0	–	–	–	–
SMT-TB	2.50	0	5.05	0	2.50	0	3.69	0	2.50	0	2.50	0	2.50	0
SMT-BB_OR_F64,4	6.17	39	8.79	132	12.83	33	6.00	46	5.79	246	**6.62**	169	6.41	92
SMT-INT	6.35	0	6.50	0	**14.61**	0	3.32	0	4.96	0	4.18	0	**10.19**	0
SMT-BB_OR	5.96	0	**18.29**	0	10.13	0	5.14	0	4.99	0	5.75	0	5.64	0
SMT-INT_OR	5.98	0	7.46	755	7.06	0	3.69	366	5.05	6	4.85	76	7.51	0
SMT-PB_OR	3.97	220	6.66	193	9.69	73	**6.74**	204	2.70	696	2.77	693	2.65	742
SMT-TB_PAV	48.29	0	7.29	0	6.07	0	5.04	0	3.05	0	3.20	0	3.06	0
SMT-TB_PAV_CC	49.10	0	7.47	0	6.21	0	5.08	0	3.23	0	3.80	0	6.54	0
SMT-TB_PAV_PAC	47.85	0	7.38	0	6.23	0	5.17	0	3.38	0	3.97	0	6.79	0

The normalized hamming distance (ham) and the number of duplicates (dupl) for the generated stimuli (up to 1,000).

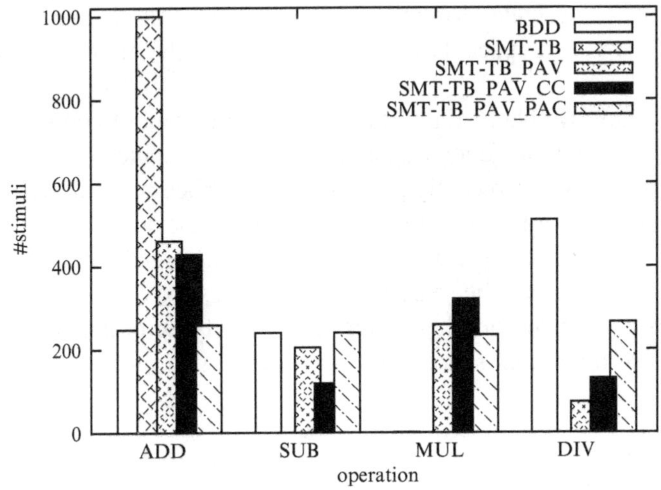

Figure 14.4. Distribution with respect to operations

Figure 14.4 shows the distribution (with respect to the stimulated operations) for the respective approaches (i.e. SMT-TB_PAV, SMT-TB_PAV_CC and SMT-TB_PAV_PAC) as well as for BDD and SMT-TB. Here, the poor distribution of SMT-TB is not surprising. But also the BDD-based approach does not lead to a uniform distribution. Thus, to achieve a uniform distribution over all operations of the device under verification, other methods must be used. In this case, the strategy SMT-TB_PAV_PAC, which incorporates the control variables, provides the best distribution of the stimuli.

Altogether, for the general case good results are obtained by applying the strategies SMT-TB_PAV, SMT-TB_PAV_CC, and SMT-TB_PAV_PAC. If additionally control path information is available, a uniform distribution over all operations can be achieved using the SMT-TB_PAV_PAC strategy.

6. Conclusions and Future Work

In this work, SMT-based stimuli generation for the SCV library has been introduced. We have shown, that using an SMT solver instead of BDDs leads to a more robust stimuli generation (in particular if complex arithmetic occurs in an SCV constraint as this is the case for many practical relevant problems). As a drawback, simply replacing the BDD engine by an SMT solver leads to poor results with respect to the distribution of constraint solutions. Thus, we proposed several strategies forcing the solver to generate stimuli from different parts of the search space. Our experiments show the effectiveness of these strategies.

In future work, we plan a tight integration of the proposed concepts into the SCV library, i.e. without generating a separate SMT instance for each stimulus. For this purpose, e.g. the recently published C-interface of Boolector [4] can be used. Furthermore, strategies directly integrated in the solve engine should be developed. More precisely, instead of excluding unwanted solutions by means of additional constraints, the search process of the solver should be influenced e.g. by adjusting the decision heuristics or by modifying propagation strategies so that parts of the search space are pruned depending on the current (partial) assignment.

Notes

1. Due to page limitation only a small set of benchmarks is presented here. However, the benchmarks include representatives of arithmetic, control logic, as well as a mixture of both and thus are sufficient to show the effects.

References

[1] A. Armando, J. Mantovani, L. Platania, Bounded model checking of software using SMT solvers instead of SAT solvers, in *SPIN*, 2006, pp. 146–162

[2] J. Bergeron, *Writing Testbenches Using SystemVerilog* (Springer, New York, 2006)

[3] M. Bozzano, R. Bruttomesso, A. Cimatti, T. Junttila, P.v. Rossum, S. Schulz, R. Sebastiani, The MathSAT 3 System, in *International Conference on Automated Deduction*, 2005

[4] R. Brummayer, A. Biere, Boolector: an efficient SMT solver for bit-vectors and arrays, in *Tools and Algorithms for the Construction and Analysis of Systems*, 2009

[5] R.E. Bryant, Graph-based algorithms for Boolean function manipulation. IEEE Trans. Comp. **35**(8), 677–691 (1986)

[6] R.E. Bryant, On the complexity of VLSI implementations and graph representations of Boolean functions with application to integer multiplication. IEEE Trans. Comp. **40**, 205–213 (1991)

[7] M. Davis, G. Logeman, D. Loveland, A machine program for theorem proving. Comm. ACM **5**, 394–397 (1962)

[8] S. Deng, J. Bian, W. Wu, X. Yang, Y. Zhao, EHSAT: an efficient rtl satisfiability solver using an extended dpll procedure, in *Design Automation Conference*, 2007, pp. 588–593

[9] B. Dutertre, L. Moura, A fast linear-arithmetic solver for DPLL(T), in *Computer Aided Verification*, volume 4114 of *LNCS*, 2006, pp. 81–94

[10] N. Eén, N. Sörensson, An extensible SAT solver, in *SAT 2003*, volume 2919 of *LNCS*, 2004, pp. 502–518

[11] V. Ganesh, D.L. Dill, A decision procedure for bit-vectors and arrays, in *Computer Aided Verification*, volume 4590, 2007, pp. 519–531

[12] D. Große, R. Ebendt, R. Drechsler, Improvements for constraint solving in the SystemC verification library, in *ACM Great Lakes Symposium on VLSI*, 2007, pp. 493–496

[13] D. Große, R. Wille, R. Siegmund, R. Drechsler, Contradiction analysis for constraint-based random simulation, in *Forum on Specification and Design Languages*, 2008, pp. 130–135

[14] IEEE 1666, *IEEE Standard SystemC Language Reference Manual*. IEEE Std. 1666, 2005

[15] C.N. Ip, S. Swan, A tutorial introduction on the new systemC verification standard. White paper, 2003

[16] H. Kim, H. Jin, K. Ravi, P. Spacek, J. Pierce B. Kurshan, F. Somenzi, Application of formal word-level analysis to constrained random simulation, in *Computer Aided Verification*, 2008, pp. 487–490

[17] N. Kitchen, A. Kuehlmann, Stimulus generation for constrainted random simulation, in *International Conference on CAD*, 2007, pp. 258–265

[18] S.K. Lahiri, R. Nieuwenhuis, A. Oliveras, SMT techniques for fast predicate abstraction, in *Computer Aided Verification*, volume 4144, 2006, pp. 424–437. Springer.

[19] J.P. Marques-Silva, K.A. Sakallah, GRASP: a search algorithm for propositional satisfiability. IEEE Trans. Comp. **48**(5), 506–521 (1999)

[20] M.W. Moskewicz, C.F. Madigan, Y. Zhao, L. Zhang, S. Malik, Chaff: engineering an efficient SAT solver, in *Design Automation Conference*, 2001, pp. 530–535

[21] S.M. Plaza, I.L. Markov, V. Bertacco, Random stimulus generation using entropy and XOR constraints, in *Design, Automation and Test in Europe*, 2008, pp. 664–669

[22] J. Rose, S. Swan, SCV randomization version 1.0., 2003

[23] SCV 1.0, *SystemC Verification Standard Specification Version 1.0e*. SystemC Verification Working Group, 2003 `http://www.systemc.org`

[24] SMT-COMP, The Satisfiability Modulo Theories Competition (SMT-COMP), 2007, `http://www.smtcomp.org`

[25] F. Somenzi, *CUDD: CU Decision Diagram Package Release 2.3.0.* University of Colorado at Boulder, 1998

[26] SystemC 2.0 *Functional Specification for SystemC 2.0.* Synopsys Inc., CoWare Inc., and Frontier Design Inc., 2002, `http://www.systemc.org`

[27] R. Wille, G. Fey, D. Große, S. Eggersglüß, R. Drechsler, Sword: a SAT like prover using word level information, in *IFIP International Conference on Very Large Scale Integration*, 2007, pp. 88–93

[28] J. Yuan, A. Aziz, C. Pixley, K. Albin, Simplifying boolean constraint solving for random simulation-vector generation. IEEE Trans CAD Integr. Circ. Syst. **23**(3), 412–420 (2004)

[29] J. Yuan, C. Pixley, A. Aziz, *Constraint-Based Verification* (Springer, 2006), New York

[30] J. Yuan, K. Shultz, C. Pixley, H. Miller, A. Aziz, Modeling design constraints and biasing in simulation using BDDs, in *International Conference on CAD*, 1999, pp. 584–590

[31] Z. Zeng, K.R. Talupuru, M.J. Ciesielski, Functional test generation based on word-level SAT. J. Syst. Archit. **51**(8), 488–511

Index

Kolibri 360 GmbH
Ferdinand-Jühlke-Straße 7
99095 Erfurt, Deutschland
produktsicherheit@kolibri360.de